WHY RELIGION MATTERS

The Fate of the Human Spirit in an Age of Disbelief

Huston Smith

HarperSanFrancisco
A Division of HarperCollins*Publishers*

HarperCollins books may be purchased for educational, business, or sales promotional use. For information please write: Special Markets Department, HarperCollins Publishers, Inc., 10 East 53rd Street, New York, NY 10022.

HarperCollins Web site: http://www.harpercollins.com
HarperCollins®, ®, and HarperSanFrancisco™ are trademarks of HarperCollins Publishers, Inc.

FIRST EDITION

Library of Congress Cataloging-in-Publication Data

Smith, Huston.
Why religion matters: the fate of the human spirit in an age of disbelief / Huston Smith.—1st ed.
p. cm.
Includes index.
ISBN 0-06-067099-1 (cloth)
ISBN 0-06-067102-5 (pbk.)
1. Religion and science. I. Title.

BL240.2.S635 2000
200—dc21 00-058188

01 02 03 04 05 RRD(H) 10 9 8 7 6

To the memory of our daughter,
Karen Senyak,
and her adopted Judaism, which sustained her in her ordeal
with sarcoma cancer;

to Antonio Banuelos,
our youngest grandchild,
who bids to see more of the third millennium
than will the rest of our family;

and for Kendra,
this love poem by the sixth Dalai Lama:

I asked so much of you
In this brief lifetime,
Perhaps we'll meet again
In the childhood of the next.

Contents

Acknowledgments

It is a pleasure to record here my gratitude to those whose assistance has made this book possible. The Rockefeller Foundation granted me an idyllic month to begin my work at its Conference and Study Center in Bellagio, Italy. Stephen Mitchell and Philip Novak have critiqued the entire typescript. Nik Warren and Jonathan Wells, respectively, have coached me where physics and biology enter the book. My editor, John Loudon, and my agent, Thomas Grady, have been friends and helpers throughout. Eric Carlson, Keith Chandler, Neal Grossman, Sarah Lewis, Robert Sprich, and Lance Trumbull have researched a number of facts and references. To all of these my sincerest thanks. I hate to think how much poorer this book would have been without their help, but none of them should be held responsible for the use I have made of their services.

My wife, Kendra, was a zealous accomplice in the last stages of revision. Years ago I told her that an important part of a wife's role is to keep her husband from making a fool of himself. Sweetly she promised, "I'll try."

PREFACE

I would place these pages last except that beginning with them may increase the chance that the reader will hear me out.

I think I have a different window onto the world, one that enables me to see things that others do not. I was born into a loving family whose parents committed their lives to the highest calling they could conceive—that of being missionaries to China. Sacrifices were to be expected, and (in the disease-ridden China of that time) they arrived on schedule; their firstborn died in their arms on his second Christmas Eve. My parents did good things. In the town they chose for their lifework there was no education for girls, so their first act was to start a girls' school. Now coeducational, it has become the most important primary school in the town.

The most important thing I inherited from my parents was faith. Its substance made me, on average, a trusting person, and its content can be stated equally simply: we are in good hands, and in gratitude for that fact we do well to bear one another's burdens.

On coming to America for college, I brought that faith with me, and the rest of my life has been a struggle to keep it intact in the face of modern winds of doctrine that assail it. If those winds were powered by truth, I would bow to them, but as I have not found them to be so, I must point that out.

Everything in this book should be read in the light of the above paragraphs, and I should also use this Preface to say a word about the book's title. The wording *Why Religion Matters* did not occur to me until after I had completed the typescript and read it straight through. The book was written on the eve of a new millennium, and to allow myself room to say what I wanted to say, I adopted as my working title "The Human Spirit in the Third Millennium." Because I knew I wanted to argue that if the human spirit is to fare better than it recently has it must shake off the tunnel vision of modernity, a working subtitle quickly suggested itself: "Light at the End of the Tunnel," with its closing punctuation—a question mark or full stop—left open.

A glance at the Table of Contents will show how resolutely I adhered to that two-part working title while writing the book, but on reading the finished product I saw that underlying and interlacing the intellectual history and social criticism that the book deals with is a pervading and urgent thesis. That thesis is the importance of the religious dimension of human life—in individuals, in societies, and in civilizations.

Rarely in these pages do I argue that importance systematically; it is better to think of the book as *demonstrating* it. To the extent that the book succeeds, it shows how and why religion matters.

Huston Smith
Berkeley, California
June 2000

Introduction

The crisis that the world finds itself in as it swings on the hinge of a new millennium is located in something deeper than particular ways of organizing political systems and economies. In different ways, the East and the West are going through a single common crisis whose cause is the spiritual condition of the modern world. That condition is characterized by loss—the loss of religious certainties and of transcendence with its larger horizons. The nature of that loss is strange but ultimately quite logical. When, with the inauguration of the scientific worldview, human beings started considering themselves the bearers of the highest meaning in the world and the measure of everything, meaning began to ebb and the stature of humanity to diminish. The world lost its human dimension, and we began to lose control of it.

The beginning of a new millennium presents itself as a fitting occasion to ponder this situation. The movements that precede millennial shifts have come and gone for this round, but before they get shelved for another thousand years, they merit a moment's reflection. The anthropologist Victor Turner suggested that these movements are to cultures what rites of passage are to individuals. They signal moments of change and transition, calling individuals and societies to connect with the symbolic roots in their past in order to prepare themselves to take the next—often frightening—step into the future. To grasp this point, we need not take the rhetoric of such movements literally. The sandwich man between placards announcing that the end is near is telling us something important, even though the end is not what he thinks it is. He is not just protesting our reigning culture. However falteringly, he is gesturing toward a heavenly city that offers an alternative to this earthly one, which is always deeply flawed.

This gives me a way to think about the book I have written, for it does indeed look back at our ancestral roots in the hope that doing so can help us understand the confusions of our present period. Cultural critics have been taking this approach for a century or more, so I owe it to the reader to explain why I have taken it upon myself to add to the library. In short, what is new here?

In a word, what is new is simplification. The danger it risks, of course, is oversimplification, and I take that risk on every page. If it be wondered where I get the courage to take on that risk, the answer is from the example of Irving Berlin. Do not laugh, for in philosophy I am he.

I will explain.

I happened to catch the *Today* show the morning after Irving Berlin died (at the age of 101, as I recall), and I was surprised to find that *Today* had invited a world-class musician, Isaac Stern, to reflect on the lifework of this tunesmith. The host of the show wanted to learn from Stern the secret of Irving Berlin's success. As a musician, Berlin was so mediocre that he could play only in the key of C, and to modulate to other keys he had to build a piano that transposed by pulling levers. Yet this run-of-the-mill musician became the most successful songwriter of the twentieth century, composing over one hundred hits, many of which will continue into the new millennium. How did Stern account for the discrepancy between Berlin's modest musical talent and his achievements?

Stern's answer was so direct that it was breathtaking. Berlin's philosophy of life (Stern proceeded to explain) was simple. He saw life as composed of a few basic elements: life and death, loneliness and love, hope and defeat—not many more. In making our way through these givens, affirmation is better than complaint, hope more viable than despair, kindness nobler than its opposite. That was about it. But because Berlin believed those platitudes implicitly, he helped people cut through the ambiguities and complexities of a confusing century.

So, piggybacking on Irving Berlin, what is obvious to me?

First, that the finitude of mundane existence cannot satisfy the human heart completely. Built into the human makeup is a longing for a "more" that the world of everyday experience cannot requite. This outreach strongly suggests the existence of the something that life reaches *for* in the way that the wings of birds point to the reality of air. Sunflowers bend in the direction of light because light exists, and people seek food because food exists. Individuals may starve, but bodies would not experience hunger if food did not exist to assuage it.

The reality that excites and fulfills the soul's longing is God by whatsoever name. Because the human mind cannot come within light-years of comprehending God's nature, we do well to follow Rainer Maria Rilke's suggestion that we think of God as a direction rather than an object. That direction is always toward the best that we can conceive, as the formula of theology's Principle of Analogical Predication indicates: when we use objects and concepts from the natural world to symbolize God, the first step is to affirm what is positive in them; the second step is to deny to God what is limiting in them; and the third step is to elevate their positive features to supereminent degree (which is to say, to the highest point that our imaginings can carry us). With God and the world categorically distinguished but nowhere disjoined, other things fall into place in the way that this book indicates.

To that metaphysical point that I find obvious, I add this historical one. Until modern science arrived, everyone lived with a worldview that conformed to the outline just mentioned. Science replaced that traditional worldview—manifold in its expressions, single in its geometrical outline—with the scientific worldview. The latest journalist to interview me remarked in the course of our conversation that I seem to be angry at science. I corrected him. I am angry at *us*—modern Westerners who, forsaking clear thinking, have allowed ourselves to become so obsessed with life's material underpinnings that we have written science a blank check. I am not talking about money here; I am talking about a blank check for science's claims concerning what constitutes knowledge and justified belief. The impressiveness of pure science enters the picture, but for the public at large the miracles of technology have generally been more important.

This is the cause of our spiritual crisis. It joins other crises as we enter the new millennium—the environmental crisis, the population explosion, the widening gulf between the rich and the poor—but those are not for this book. It now remains only to outline the course the book will take.

Chapter One backs up for a running start and traces the three historical periods that have brought us to where we now are, highlighting the accomplishments and deficiencies of each. Chapter Two describes the spiritual dimensions of the world that people inhabited before being shunted by our *misreading* of modern science—I emphasize that the culprit is not science itself but our misconstrual of it—into the tunnel that serves as the presiding metaphor for this book. (Its resemblance to Plato's cave will not be lost on readers.) Descriptions of the tunnel and its four sides occupy Part One.

Part Two looks to the future as symbolized by light at the tunnel's end. Its first several chapters toy with predictions, but it then settles into its main job, which is to describe features of the religious landscape that are invariant. The strategy is straightforward. Because prediction is a hazardous enterprise, it yields diminishing returns. The best way to prepare for the future is to have in our possession a map that can orient us, wherever the future may bring.

PART 1

MODERNITY'S TUNNEL

I move into Part One of this book by way of three quotations. They are longer than I might wish, but there is good reason to quote them in full. For whatever the reader may think of the controversial chapters that follow, I do not see how (after reading these quotations) it will be possible to doubt that the chapters are set on sound foundations.

The first quotation is by a colleague of mine while I was teaching at Syracuse University, the sociologist Manfred Stanley.

> *It is by now a Sunday-supplement commonplace that the . . . modernization of the world is accompanied by a spiritual malaise that has come to be called alienation. . . . At its most fundamental level, the diagnosis of alienation is based on the view that modernization forces upon us a world that, although baptized as real by science, is denuded of all humanly recognizable qualities; beauty and ugliness, love and hate, passion and fulfillment, salvation and damnation. It is not, of course being claimed that such matters are not part of the existential realities of human life. It is rather that the scientific worldview makes it illegitimate to speak of them as being "objec-*

> *tively" part of the world, forcing us instead to define such evaluation and such emotional experience as "merely subjective" projections of people's inner lives.*

Ernest Gellner, sociologist and philosopher, picks up where Stanley leaves off by admitting that we have no reason to think that the world in itself is as Stanley describes it. It is just that we are now constrained to think that that is its character because the Promethean concerns that power the modern world decree that the only "true" knowledge is the kind that digs into nature's foundations and increases our ability to control it. In Gellner's words:

> *It was Kant's merit to see that this [epistemological] compulsion is in us, not in things. It was Weber's to see that it is historically a specific kind of mind, not human mind as such, which is subject to this compulsion. . . . We have become habituated to and dependent on effective knowledge [as just described] and hence have bound ourselves to this kind of genuine explanation. . . . "Reductionism," the view that everything in the world is really something else, and that something else is coldly impersonal, is simply the ineluctable corollary of effective explanation.*

Gellner admits that this epistemology that our Prometheanism has forced upon us carries morally disturbing consequences:

> *It was also Kant's merit to see the inescapable price of this Faustian purchase of real [sic] knowledge. [In delivering] cognitive effectiveness [it] exacts its inherent moral, "dehumanizing" price. . . . The price of real knowledge is that our identities, freedom, norms, are no longer underwritten by our vision and comprehension of things. On the contrary we are doomed to suffer from a tension between cognition and identity.*

Hannah Arendt carries these thoughts to their metaphysical conclusion:

What has come to an end is the distinction between the sensual and the supersensual, together with the notion, at least as old as Parmenides, that whatever is not given to the senses . . . is more real, more truthful, more meaningful than what appears; that it is not just beyond sense perception but above the world of the senses. . . . In increasingly strident voices, the few defenders of metaphysics have warned us of the danger of nihilism inherent in this development. The sensual . . . cannot survive the death of the supersensual [without nihilism moving in].

With these thoughts clearly before us, welcome to the tunnel of modernity.

CHAPTER 1

WHO'S RIGHT ABOUT REALITY: TRADITIONALISTS, MODERNISTS, OR THE POSTMODERNS?

Wherever people live, whenever they live, they find themselves faced with three inescapable problems: how to win food and shelter from their natural environment (the problem nature poses), how to get along with one another (the social problem), and how to relate themselves to the total scheme of things (the religious problem). If this third issue seems less important than the other two, we should remind ourselves that religious artifacts are the oldest that archeologists have discovered.

The three problems are obvious, but they become interesting when we align them with the three major periods in human history: the traditional period (which extended from human beginnings up to the rise of modern science), the modern period (which took over from there and continued through the first half of the twentieth century), and postmodernism (which Nietzsche anticipated, but which waited for the second half of the twentieth century to take hold).

Each of these periods poured more of its energies into, and did better by, one of life's inescapable problems than did the other two. Specifically, modernity gave us our view of *nature*—it continues to be refined, but because modernity laid the foundations for the scientific understanding of it, it deserves credit for the discovery.

Postmodernism is tackling *social injustices* more resolutely than people previously did. This leaves *worldviews*—metaphysics as distinct from cosmology, which restricts itself to the empirical universe—for our ancestors, whose accomplishments on that front have not been improved upon.

The just-entered distinction between cosmology and metaphysics is important for this book, so I shall expand it slightly. *Cosmology* is the study of the physical universe—or the world of nature as science conceives of it—and is the domain of science. *Metaphysics,* on the other hand, deals with all there is. (The terms *worldview* and *Big Picture* are used interchangeably with *metaphysics* in this book.) In the worldview that holds that nature is all there is, metaphysics coincides with cosmology. That metaphysics is named *naturalism.*

Such is the historical framework in which this book is set, and the object of this chapter is to spell out that framework. Because I want to proceed topically—from nature, through society, to the Big Picture, tying each topic to the period that did best by it—this introduction shuffles the historical sequence of the periods. I take up modernity first, then postmodernity, leaving the traditional period for last.

Modernity's Cosmological Achievement

In the sixteenth and seventeenth centuries Europe stumbled on a new way of knowing that we refer to as the *scientific method.* It centers in the controlled experiment and has given us modern science. Generic science (which consists of careful attention to nature and its regularities) is as old as the hills—at least as old as art and religion. What the controlled experiment adds to generic science is proof. True hypotheses can be separated from false ones, and brick by brick an edifice has been erected from those proven truths. We commonly call that edifice the *scientific worldview,* but *scientific cosmology* is more precise because of the ambiguity of the word *world.* The scientific edifice is a *world*view only for those who assume that science can in principle take in all that exists.

The scientific cosmology is so much a part of the air we breathe that it is hardly necessary to describe it, but I will give it a paragraph to provide a reference point for what we are talking about. Some fifteen billion years ago an incredibly compact pellet of matter exploded to launch its components on a voyage that still continues. Differentiation set in as hydrogen proliferated into the periodic table. Atoms gathered into gaseous clouds. Stars condensed from whirling filaments of flame, and planets spun off from those to become molten drops that pulsated and grew rock-encrusted. Narrowing our gaze to the planet that was to become our home, we watch it grow, ocean-filmed and swathed in atmosphere. Some three and a half billion years ago shallow waters began to ferment with life, which could maintain its inner milieu through homeostasis and could reproduce itself. Life spread from oceans across continents, and intelligence appeared. Several million years ago our ancestors arrived. It is difficult to say exactly when, for every few years paleontologists announce discoveries that "set the human race back another million years or so," as press reports like to break the news.

Taught from primary schools onward, this story is so familiar that further details would only clutter things.

Tradition's Cosmological Shortcomings

That this scientific cosmology retires traditional ones with their six days of creation and the like goes without saying. Who can possibly question that when the scientific cosmology has landed people on the moon? Our ancestors were impressive astronomers, and we can honor them unreservedly for how much they learned about nature with only their unaided senses to work with. And there is another point. There is a naturalism in Taoism, Zen Buddhism, and tribal outlooks that in its own way rivals science's calculative cosmology, but that is the naturalism of the artist, the poet, and the nature lover—of Li Po, Wordsworth, and Thoreau, not that of Galileo and Bacon. For present purposes, aesthetics is irrelevant. Modern cosmology derives from laboratory experiments, not landscape paintings.

Postmoderism's Cosmological Shortcomings

With traditional cosmology out of the running, the question turns to postmodernism. Because science is cumulative, it follows as a matter of course that the cosmology we have in the twenty-first century is an improvement over what we had in the middle of the twentieth, which on my timeline is when modernity phased into postmodernity. But the refinements that postmodern scientists have achieved have not affected life to anything like the degree that postmodern social thrusts have, so the social Oscar is the one postmodernists are most entitled to.

The next section of this chapter will discuss postmodernism's achievements on the social front, but before turning to those I need to support my contention that postmodern science (it is well to say postmodern *physics* here) does not measure up to modern physics in the scope of its discoveries. It says nothing against the brilliance of Stephen Hawking, Fred Hoyle, John Wheeler, Freeman Dyson, Steven Weinberg, and their likes to add that they have discovered nothing about nature that compares with the discoveries of Copernicus, Newton, Maxwell, Planck, Einstein, Heisenberg, Bohr, Schrödinger, and Born. In molecular chemistry things are different. DNA is a staggering discovery, but, extending back only millions of years compared with the astrophysicists' billion, it is not in nature's foundations. The fact that no new abstract idea in physics has emerged for seventy years may suggest that nothing more remains to be discovered about nature's foundations. Be that as it may, postmodernism's discoveries (unlike modern discoveries in physics—the laws of gravity, thermodynamics, electromagnetism, relativity theory, and quantum mechanics, which continue to be used to make space shuttles fly and to help us understand how hot electrons behave in semiconductors) have concerned details and exotica. The billions of dollars that have been spent since the middle of the twentieth century (and the millions of papers that have been written on theories that change back and forth) have produced no discoveries that impact human beings in important ways. All are in the domain of the metasciences of high-energy particle physics and astronomy, whose findings—what is supposed to have happened in the first 10^{-42} seconds of the universe's life, and the like—while headlined by the media have no conceivable connection to human life and can be neither falsified nor checked in normal ways. This allows the building blocks of nature—particles, strings, or whatever—to keep changing, and the age of the universe to be halved or doubled every now and then. Roughly 99.999 percent of science (scientist Rustum Roy's estimate) is unaffected by these flickering hypotheses, and the public does not much care about their fate.

Outranking the foregoing reason for not giving the cosmological Oscar to postmodernism is the fact that the noisiest postmodernists have called into question the very notion of truth by turning claims to truth into little more than power plays. According to this reading of the matter, when people claim that what they say is true, all they are really doing is claiming status for beliefs that advance their own social standing. This relativizes science's assertions radically and rules out even the possibility of its closing in on the nature of nature. The most widely used textbook on college campuses for the past thirty years has been Thomas Kuhn's *The Structure of Scientific Revolutions,* and its thesis—that facts derive their meaning from the paradigms that set them in place—has shifted attention from scientific facts to scientific paradigms. As there are no neutral standards by which to judge these paradigms, Kuhn's thesis (if unnuanced) leads to a relativism among paradigms that places Hottentot science on a par with Newton's. Kuhn himself phrased his thesis carefully enough to parry such relativism, but even taken at its best, it provides no way that science could get to the bottom of things. This demotes the enterprise, and in doing so provides a strong supporting reason for not giving postmodernism the cosmological prize. It does better with social issues, to which I now turn.

Postmodernism's Fairness Revolution

The magic word of postmodernism is *society*. This is not surprising. With the belief that there is nothing beyond our present world, nature and society are all that remain, and of the two nature has become the province of specialists. We seldom confront it directly anymore; mostly it comes to us via supermarkets and cushioned by air-conditioning and central heating. This leaves society as the domain that presses on us directly and the one in which there is some prospect of our making a difference.

And changes are occurring. Post colonial guilt may play a part here, and so much remains to be done that self-congratulation is premature. Still, a quick rehearsal of some changes that have occurred in a single lifetime makes it clear that social injustices are being recognized and addressed more earnestly today than they were by our ancestors:

- In 1919 the Brooklyn Zoo exhibited an African American caged alongside chimpanzees and gorillas. Today such an act would be met with outrage anywhere in the world.
- The civil rights movement of the 1960s accomplished its major objectives. In the United States and even in South Africa today, people of different races mix where they never could before—on beaches, in airline cabin crews, everywhere.
- In the 1930s, if a streetcar in San Francisco approached a stop where only Chinese Americans were waiting to board, it would routinely pass them by. By contrast, when (fifty years later) I retired from teaching at the University of California, Berkeley, my highly respected chancellor was a Chinese American who spoke English with a Chinese accent.
- No war has ever been as vigorously protested as was the war in Vietnam by United States citizens. When things were going so badly that military leaders advised President Nixon to use nuclear weapons, he declined because (as he said) if he did that, he would face a nation that had taken to the streets.
- The women's movement is only a blink in the eyes of history, but it has already scored impressive victories. Until long after the Civil War, American women really had no civil rights, no legal rights, and no property rights. Not until 1918 did Texas alter its law that everyone had the right to vote except "idiots, imbeciles, aliens, the insane, and women."
- Arguably, the most important theological development of the latter twentieth century was the emergence of the theology of liberation, with its Latin American and feminist versions in the vanguard.
- In an unprecedented move, in March 2000 the pope prayed to God to forgive the sins his church had committed against the people of Israel, against love, peace, and respect for cultures and religions, against the dignity of women and the unity of the human race, and against the fundamental rights

of persons. Two months later, two hundred thousand Australians marched across Sydney Harbor Bridge to apologize for their treatment of the aborigines while the skywritten word SORRY floated above the Sydney Opera House.

Tradition's Social Shortcomings

These signs of progress acquire additional life when they are set against the unconcern of earlier times regarding such matters. There is no reason to think that traditional peoples were more callous than we are, but on the whole they saw their obligations as extending no further than to members of their primary communities: Buddhism's *dana* (gifts), Jesus' "cup of water given in my name," and their likes. Encountered face-to-face, the hungry were fed, the naked were clothed, and widows and orphans were provided for as means allowed, but there human obligations ended. Injustices that were built into institutions (if such injustices were even recognized) were not human beings' responsibility, for those institutions were considered to be God-given and unalterable. People regarded them in the way we regard laws of nature—as givens to be worked with, not criticized.

Modernity changed this attitude. Accelerating travel and trade brought encounters between peoples whose societal structures were very different from one another, and these differences showed that such institutions were not like natural laws after all; they were humanly devised and could therefore be critiqued. The French Revolution put this prospect to a historic test; scrapping the divine right of kings, it set out to create a society built on liberty, equality, and fraternity. The experiment failed and the backlash was immediate, but its premise—that societies are malleable—survived.

Modernity's Social Shortcomings

Modernity deserves credit for that discovery, and (if we wished) we might excuse it for its poor handling of its discovery on grounds that it was working with a new idea. The record itself, however, is by postmodern standards, deplorable. Under the pretext of shouldering "the white man's burden" to minister to "lesser breeds without the law," it ensconced colonialism, which raped Asia and Africa, hit its nadir in the Opium Wars of 1841–42, and ended by subjecting the entire civilized world to Western domination. David Hume is commonly credited with having the clearest head of all the great philosophers, but I read that somewhere in his correspondence (I have not been able to find the passage) he wrote that the worst white man is better than the best black man. What I can report firsthand is signs posted in parks of the international settlements in Shanghai, where I attended high school, that read, "No dogs or Chinese allowed." With a virgin continent to rape, the United States did not need colonies, but this did not keep it from hunting down the Native Americans, continuing the institution of slavery, annexing Puerto Rico and Hawaii, and establishing "protectorates" in the Philippines and several other places.

Having dealt with nature and society, I turn now to the third inescapable issue that human beings must face: the Big Picture.

THE TRADITIONAL WORLDVIEW

Modernity's Metaphyscial's Shortcomings

Modernity was metaphysically sloppy. Ravished by science's accomplishments, it elevated the scientific method to "our sacral mode of knowing" (Alex Comfort), and because that mode registers nothing that is without a material component, immaterial realities at first dropped from view and then (as the position hardened) were denied existence. In the distinction registered earlier in this chapter, this was metaphysics reduced to cosmology. When Carl Sagan opened his television series, *Cosmos,* by announcing that "the Cosmos is all that is or ever was or ever will be," he presented that unargued assumption as if it were a scientific fact. Modernity's Big Picture is materialism or (in its more plausible version) naturalism, which acknowledges that there *are* immaterial things—thoughts and feelings, for example—while insisting that those things are totally dependent on matter. Both versions are stunted when compared with the traditional outlook. It is important to understand that neither materialism nor naturalism is required by anything science has discovered in the way of actual facts. We have slid into this smallest of metaphysical positions for psychological, not logical, reasons.

Postmodernity's Metaphyscial Shortcomings

As for postmodernity, it sets itself against the very idea of such a thing as the Big Picture. It got off on the right foot by critiquing the truncated worldview of the Enlightenment, but from that reasonable beginning it plunged on to argue unreasonably that worldviews (often derisively referred to as *grand narratives*) are misguided in principle. In *The Postmodern Condition,* Jean François Lyotard goes so far as to *define* postmodernism as "incredulity toward metanarratives," a synonym for metaphysics.

The incredulity takes three forms that grow increasingly shrill as they proceed. Postmodern *minimalism* contents itself with pointing out that we have no consensual worldview today; "we have no maps and don't know how to make them." *Mainline* postmodernism adds, "and never again will we have a consensual worldview, such as prevailed in the Middle Ages, Elizabethan England, or seventeenth-century New England; we now know too well how little the human mind can know." *Hardcore* postmodernism carries this trajectory to its logical limit by adding, "good riddance!" Stated in the in-house idiom postmodernists are fond of, worldviews "totalize" by "marginalizing" minority viewpoints. They are oppressive in principle and should be resolutely resisted.

If hardcore postmodernism were accurate in this charge it would stop this book in its tracks, but it has not proved that it is accurate—it merely *assumes* that it is accurate and rests its case on examples of oppression that, of course, are not lacking. What has not been demonstrated is the impossibility of a worldview that builds the rights of minorities into its foundations as an essential building block. There is irony here, for the very postmodernism that is dismissing the possibility of a comprehensive humane outlook is working toward the creation of such through its fairness revolution—its insistence that everybody be given an equal chance at the goods of life. The deeper fact, however, is that to have or not have a worldview is not an option, for peripheral vision always conditions what we are attending to focally, and in conceptual "seeing" the periphery has no cutoff. The only choice we have is to be consciously aware of our worldviews and criticize them where they need criticizing, or let them work on us unnoticed and acquiesce to living unexamined lives.

To say as I have that neither modernism nor postmodernism handled the metaphysical problem well is, of course, no proof that traditionalists handled it better. If this chapter were a self-contained unit, I would need to complete it by describing the traditional worldview and defending its merits. That, however, comes so close to being the object of this entire book that I will not try to compress it into a page or two here. Moreover, the traditional worldview is so out of favor today that the only possible way to gain a hearing for it is to ease into it, so to speak, by suggesting plausibilities wherever openings for them appear.

That leaves this present chapter open ended, but even so these early pages have accomplished two things. The first of these is descriptive: this chapter has placed the present in its historical setting. The second is prescriptive, for an obvious moral emerges from what has been said. We should enter our new millennium by running a strainer through our past to lift from each of its three periods the gold it contains and let its dross sink back into the sands of history. Modernity's gold—i.e., science—is certain to figure importantly in the third millennium, and postmodernity's focus on justice likewise stands a good chance of continuing. It is the traditional worldview that is in jeopardy and must be rehabilitated if it is to survive.

CHAPTER 2

The Great Outdoors and the Tunnel Within It

Those who fear that my announced regard for the traditional worldview may mean that this book is headed toward a nostalgia trip can rest assured; there will be no pining here for the so-called good old days. When the Irish joke, "To hell with the future; we live in the past," I enjoy the wry humor, but it is no stance for life. In using a tunnel as my presiding metaphor for this book, I am not arguing that as a whole our times are worse than others. Things look scary in many ways, but there are encouraging signs as well, and to come right to the point, the entire issue of historical comparisons is beyond me. I have neither taste nor talent for the project. I have no idea what it felt like to live in ancient times. When? Where? In which sex and what class? In order even to get to those questions one needs a standard, and I have no way of determining what life *on average* feels like even today. Who do we include in our sample? AIDS-ridden Africans? New Zealand shepherds? Affluent CEOs who enjoy salaries four hundred times the wages of their employees? Inner-city African Americans and the homeless of all sorts? Mix all of these, shake well, and what quality of life emerges?

And as if these imponderables were not already sufficient, presiding over them all is the mood one happens to be in when one takes the reading. Nobel prize–winning poet Czeslaw Milosz tells us that "on one side there is luminosity, trust, faith, the beauty of the earth; on the other side, darkness, doubt, unbelief, the cruelty of the earth, the capacity of people to do evil. When I write, the first side is true; when I do not write, the second is." Touché! I myself regularly receive letters both from doomsday prophets who see us going down the drain like Rome and from their opposite numbers—bright-eyed, bushy-tailed New Agers who sound as if they expect a mutation of consciousness to reopen the gates of Eden for two-way traffic any day now. I wish I could readdress each letter, unopened, to one of its opposite number. Let them fight it out while I hold their jackets. The French adage, *Plus ça change, plus c'est la même chose* ("The more things change, the more they stay the same") rings truer to me, as does the opening line of Charles Dickens's *A Tale of Two Cities:* "It was the best of times, it was the worst of times."

Still, I cannot side completely with those sentences either, for if I did there would be no point in writing this book: it could make no difference. So I must backpedal a bit.

Improvements are possible and we should do our best to effect them, but as Robert Frost put the matter, each must wreak our will on the world in our own way. My way relates to worldviews. I am convinced that whatever transpires in other domains of life—politics, living standards, environmental conditions, interpersonal relationships, the arts—we will be better off if we extricate ourselves from the worldview we have unwittingly slipped into and replace it with a more generous and accurate one. That, and that only, is the concern of this book. Naturally, I hope for the best on life's other fronts and do what I can to better them, but there is nothing I might say regarding them that warrants the reader's attention. With worldviews the case is different. On and off, background and foreground, I have been pondering the ultimate nature of things since as far back as I can remember, and I would have to think poorly of myself—suffer from "a low self of steam," as a high school student wrote—if the sheer quantity of that cerebration had not produced something worth contributing to the conversation.

Worldviews: The Big Picture

This focus for the book poses a problem, for metaphysics is not an eye-catching subject. (As I have said, I use the terms *metaphysics, worldview,* and *Big Picture* interchangeably.) It resembles peripheral vision, which gets overlooked precisely for *being* peripheral. That does not render it unimportant, however. Psychologists tell us that what we are looking at is always affected by its background, and that holds as much for the way the world appears and feels to us as it does for ocular vision. Traditional people sensed this, and Claude Levi-Strauss was perceptive in picking up that point; the deepest difference between their epistemology and ours, he concluded, was that "primitives" believe that you cannot understand anything unless you understand everything. He thought they were wrong in so thinking, but when we enter the background-foreground issue we see that they had things exactly right. With us, life's problems press so heavily on us that we seldom take time to reflect on the way our unconscious attitudes and assumptions about the nature of things affect the way we perceive what is directly before us. It takes philosophers to call attention to this oversight, as William James did when he said that in interviewing a prospective roomer a landlady would do better to inquire about his philosophy of life than to check his bank account. John Stuart Mill tied that point directly to metaphysics: "If it were not useful to know in what order of things, under what government of the universe it is our destiny to live," he wrote, "it is difficult to imagine what could be considered so, for whether a person is in a pleasant or in an unpleasant place, a palace or a prison, it cannot be otherwise than useful to him to know where he is." The nineteenth-century zoologist Ernst Haeckel said that if he could have one question answered authoritatively it would be, Is the universe friendly?

Those reports tell the story, but they are one-liners, so I will flesh them out slightly.

A generation ago, psychologist William Sheldon of Columbia University's College of Physicians and Surgeons wrote that "continued observations in clinical practice lead almost inevitably to the conclusion that deeper and more fundamental than sexuality, deeper than the craving for social power, deeper even than the desire for possessions, there is a still more generalized and universal craving in the human makeup. It is the craving for knowledge of the right direction—for orientation." Such orientation requires knowing the lay of the land if only intuitively, and there is no cutoff point for the geography in question.

Stated in my own words, the point comes down to this: minds require eco-niches as much as organisms do, and the mind's eco-niche is its worldview, its sense of the whole of things (however much or little that sense is articulated). Short of madness, there is *some* fit between the two, and we constantly try to improve that fit. Signs of a poor fit are the sense of meaninglessness, alienation, and anxiety that the twentieth century knew so well. The proof of a good fit is that life and the world make sense. When the fit feels perfect, the energies of the cosmos pour into the believer and empower her to a startling degree. She knows that she belongs. The Ultimate supports her, and the knowledge that it does that produces a wholeness that is solid for fitting as a piece of a jigsaw puzzle into the wholeness of the All.

If these few sentences have not convinced the reader of the importance of metaphysics, I doubt that anything else I could say would do the job, so I shall proceed from metaphysics in general to the contrasting worldviews that concern this book—the traditional and the scientific Big Pictures.

THE DECISIVE ALTERNATIVE

Two working principles control this comparison. The first is Max Weber's notion of ideal types. Ideal types are like platonic forms or mathematical lines. Never perfectly instantiated in our imperfect world, they can (nevertheless) serve as heuristic devices to help us keep our ideas in order. My second strategy is to table in this chapter the question of truth: which of the two Big Pictures, traditional and scientific, do we have reason to believe coincides most closely with the nature of things? That question supersedes everything else I shall be saying about the worldviews, but I am reserving it for later, where it will be addressed head on. Here at the beginning I am concerned with which view is descriptively superior. If we had our choice, which would we prefer?

The Enchanted Garden

In describing the traditional world as an enchanted garden, Max Weber created a lovely if wistful metaphor. He used the designation condescendingly, for the enchantment he had in mind was that of children who experience the world in its original freshness, unmarred by habituation—in Wordsworth's words, the "time when meadow, grove and stream, the earth and every common sight, to me did seem apparelled in celestial light, the glory and the freshness of a dream." (A friend of mine told me that after he outgrew that childhood stage, he could recover its magic by bending over and viewing the world upside down through his outspread legs. The technique worked for less than a year, however.) Weber's Enlightenment belief that early peoples were children in comparison with ourselves has collapsed, but as a freestanding metaphor the enchanted garden is appropriate, for (unconfined by the tunnel that I shall be coming to) past peoples had vistas to look out upon. My variant of Weber's garden, the great outdoors, allows room for those vistas. Some of them are terrifying, but that does not affect their majesty and expanse.

I will speak more plainly. The traditional worldview is preferable to the one that now encloses us because it allows for the fulfillment of the basic longing that lies in the depths of the human heart. I mentioned that longing in the Introduction and need now to describe it more fully.

There is within us—in even the blithest, most lighthearted among us—a fundamental dis-ease. It acts like an unquenchable fire that renders the vast majority of us incapable in this life of ever coming to full peace. This desire lies in the marrow of our bones and the deep regions of our souls. All great literature, poetry, art, philosophy, psychology, and religion tries to name and analyze this longing. We are seldom in direct touch with it, and indeed the modern world seems set on preventing us from *getting* in touch with it by covering it with an unending phantasmagoria of entertainments, obsessions, addictions, and distractions of every sort. But the longing is there, built into us like a jack-in-the-box that presses for release. Two great paintings suggest this longing in their titles—Gauguin's *Who Are We? Where Did We Come From? Where Are We Going?* and de Chirico's *Nostalgia for the Infinite*—but I must work with words. Whether we realize it or not, simply to be human is to long for release from mundane existence, with its confining walls of finitude and mortality.

Release from those walls calls for space outside them, and the traditional world provides that space in abundance. It has about it the feel of long, open distances and limitless vistas for the human spirit to explore—distances and vistas that are quality-laden throughout. Some of its vistas, as I mentioned, are terrifying; still, standing as it does as the qualitative counterpart to the quantitative universe that physics explores, all but the fainthearted would switch to it instantly if we believed it existed (so at least this chapter argues). Our received wisdom denies its existence, but that wisdom cannot prevent us from having experiences that *feel* as if they come from a different world.

Mystics are people who have a talent for sensing places where life's carapace is cracked, and through its chinks they catch glimpses of a world beyond. Isaiah seeing the Lord high and lifted up. Christ seeing the heavens open at his baptism. Arjuna privy to Krishna in his terrifying cosmic form. The Buddha finding the universe turning into a bouquest of flowers at the hour of his enlightenment. John reporting, "I was on an island called Patmos, and I was in a trance." Saul struck blind on the Damascus road. For Augustine it was the voice of a child, saying, "Take, read"; for Saint Francis, a voice that seemed to come from the crucifix. It was while Saint Ignatius sat by a stream and watched its running water, and while that curious old cobbler Jacob Boehme was looking at a pewter dish, that there came to each that news of another world that it is always religion's business to convey.

Stories grow up around theophanies such as these, and in the course of generations they condense into myths that impregnate cultures with meaning and motivation. Science provides a useful analogy here. The entire scientific worldview has been spun from a relatively few crucial experiments, which can be likened to the numbered dots in children's puzzles that (when they are connected by a line that is drawn through them sequentially) produces the outline of a giraffe or whatever. Myths are like the lines traditional peoples collectively and largely unconsciously draw to connect the "dots" of the direct disclosures that their visionaries report.

If number is the language of science, myth is the language of religion. It does not map literally onto the commonsense world—biblical literalists' mistake is to think that it does—but that is not a problem, for as Steven Weinberg tells us, "We know how hopeless it is trying to fit quantum mechanics [too] into our everyday world." The signature of myth is always its happy ending, which makes myths like fairytales writ large. Fairytales locate their happy endings—marrying the princess—in this world; myths anchor that ending in the final nature of things, which conquers death itself. It is the most successful plot device that has ever been conceived, and it is easy to see why, for it stretches our imaginations to their limits and goes on from there to assure us that we can have what they reach for. But only *after,* let it never be forgotten, we have faced and surmounted ordeals of daunting magnitude. Philo of Alexandria and Origen (to name only two) called this plot device the Principle of Maximum Meaning and proposed it as the ruling principle for all exegesis. Does the scriptural passage in question inspire and strengthen us?

I have been considering the great outdoors from the human standpoint by trying to suggest how it might feel to live in it, but the decisive thing about it has yet to be mentioned. Traditional peoples do not think of the tangible world as freestanding. It derives from a divine source, called the Great Spirit, God, the One, the Infinite, whatever. This source is not separate from the world—separation is the only thing it is separate *from.* It is, however, exempt from the world's limitations: time with its perpetual perishings, space with its separations, and finitude with its oppressive restrictions. Our forebears took this distinction between Transcendence and immanence (the former capitalized to indicate its superiority) as a summons to return to the source from which they derived. "God became man that man might become God" is the way Christians put the matter. The Buddhist version reads, "There is an unborn, an unbecome, an unmade, an uncompounded, therefore there is an escape from the born, the become, the made, the compounded."

To say that the pilgrim is not alone in her heroic journey understates the case, for it is the spark of divinity that God plants in human beings that initiates the journey in the first place. Transcendence takes the initiative at every turn: in creating the world, in instantiating itself in the world, and in shaping civilizations through its revelations—revelations that set civilizations in motion and establish their trajectories. This is the invincible ground for hope that writes optimism into the traditional worldview. That the divine *must* take the initiative if the world's return to it is to succeed is evident to anyone with the slightest metaphysical flair, for the incommensurable difference between the finite and the infinite renders absurd the notion that the finite might bridge the gap with its own resources. If there is to be a wooing of God by humanity, it must be God who is the real agent in the wooing, as well as its object.

I am trying to keep to my resolve to reserve questions of truth for later chapters, but being myself a child of our skeptical age, I find charges of "wishful thinking," "escapism," and "hopes for peace of mind" gnawing around the edges of these paragraphs as I write them, so I will interrupt for a moment to face them, beginning with a fourth charge that was not on my original list because I had not yet gotten around to the sociobiologist E. O. Wilson's *Consilience*. When I did read it, I discovered on page 286 that people follow religion because it is "easier" than empiricism.

That struck a nerve, and provoked a response I shall be candid enough to report. Mr. Wilson:

- When you have endured an eight-day *O-sesshin* in a Zen monastery, sitting cross-legged and motionless for twelve hours a day and allowed only three and one-half hours of sleep each night until sleep and dream deprivation bring on a temporary psychosis (my own nondescript self);
- When you have attended four "rains retreats" at the Insight Buddhist Meditation Center in Barre, Massachusetts, for a total of one complete year of no reading, no writing, no speaking, and eyes always downcast (my wife);
- When you have almost died from the austerities you underwent before you attained enlightenment under a bo tree in India;
- When you have been crucified on Golgatha;
- When you have been thrown to lions in the Roman coliseum;
- When you have been in a concentration camp and held on to some measure of dignity through your faith;
- When you have given your life to providing a dignified death for homeless, destitute women gathered from the streets of Calcutta (Mother Teresa), or played out her counterpart with the poor in New York City (Dorothy Day);

When, Mr. Wilson, you have undergone any one of these trials, it will then be time to talk about the ease of religion as compared with the ardors of empiricism.

That outburst weathered, I proceed with the escapism I was going to address before I lost my temper. There are times when efforts to escape are not ignoble. Why should a man be scorned if, finding himself in prison, he tries to get out and go home? My mind goes to something Walter Capps (the only professor of religion to have been elected to the U.S. Congress) told me shortly before his life was cut short by a fatal heart attack. His scholarly work, which ranged widely, included a book on monasticism. As part of his research for that book he paid a number of live-in visits to monasteries. As he was driving home from one of them, preoccupied with the question of whether monks were escapists, he noticed a department store and stopped to pick up an item that he needed. It was early in the morning (monks get up early) and he found himself in a crowd of women waiting for the store to open its doors, which it did a moment or two after his arrival. It turned out be the opening of a giant lingerie sale, so Walter was swept into the store by a river of women who plunged for its mountains of underwear and began pawing through them frantically to get their pick of the bargains. The spectacle, he told me, threw light on the question that had been troubling him. Was it the monks he had left an hour earlier who were the escapists, or those bargain-seekers who looked as if they were trying to assuage their spiritual emptiness with cut-rate underwear?

The Tunnel

The arrival of modern science has had consequences that our ancestors could not have remotely envisioned. That science has changed our world beyond recognition goes without saying, but it is the way that it has changed our worldview that concerns this book. The traditional worldview has fallen before it, but not completely. Historical changes do not turn on a knife-edge, for tradition pulls against them, making cultural lag a factor that must always be reckoned with. This is why (as I earlier mentioned) I am resorting to Max Weber's ideal types. Very few, if any, people today subscribe to either the scientific or the traditional worldview without (unconsciously if not consciously) smuggling in some features of the other outlook. Even those who have abandoned the theological specifics of the religious view continue to linger in its afterglow by believing that human beings are endowed with certain unique properties (inherent dignity and inalienable rights), that other organisms do not possess, and that (as a consequence) the highest priority a democratic society can set for itself is to respect the sanctity and worth of the individual. Evolutionists try to justify these tenets as emergent values, but what passes unnoticed (as Walker Percy pointed out) is that the moment the sanctity of the individual is turned into a "value," a huge act of devaluation has already occured.

If we add to the fact that no one subscribes to either the scientific or the traditional worldview in its uncontaminated form the further fact that no two persons see either view in exactly the same way, my reason for presenting the views as ideal types becomes clear.

Weighing the Alternatives

To etch the alternatives in sharp relief, I shall mention five places where they contradict each other.

1. In the traditional, religious view *spirit is fundamental and matter derivative.* (In this discussion, unless otherwise indicated, I use the terms *traditional* and *religious* interchangeably, for all traditional societies were religious.) Matter obtrudes in the sea of spirit only occasionally, like icebergs. The scientific worldview turns this picture on its head. In restricting consciousness (which is as close to spirit as science gets) to attributes of complex organisms, it turns spirit into tiny rivulets on a single planet in a desert approximately fifteen billion light-years across.

2. In the religious worldview human beings are *the less who have derived from the more.* Trailing clouds of glory, they carry within themselves traces of their noble origins. They are creatures of their Creator, or (stated philosophically) emanations from the One that

contains every perfection. Tribal peoples couch the point graphically, as when the Tukano people of the Vaupés region of Colombia say that the first people came from the sky in a serpent canoe. Science reverses this etiology, positioning humanity as *the more that has derived from the less.* From a universe that was devoid of sentience at its start, life eventually emerged, and from its simplest form has advanced to the elevated stature that we humans beings now enjoy. Nothing in science's universe is more intelligent than we are.

3. The traditional worldview points toward a *happy ending;* the scientific worldview does not. In the Abrahamic family (which is more invested in history than tribal and Asian religions are) both individual souls and history as a whole end happily. History climaxes in the coming of the Messiah (Judaism), the Second Coming of Christ (Christianity), and the coming of *al-Mahdi,* the Rightly Guided, who appears before the end of time and restores the ties between heaven and earth until time shuts down (Islam). Heaven provides the happy ending for individual souls, but the doctrine of eternal damnation in hell obtrudes glarringly here—so much so that I will devote a paragraph to it after I complete this Cook's Tour of final things. Unlike their Western counterparts, India's religions consider *samsara* (roughly the world we now know) to be unredeemable, but Hinduism and Buddhism, along with India's less populous religions, all teach the possibility of escape from *samsara* into *Nirvana.* East Asia (recognizing itself to be metaphysically mediocre and not much interested in the subject) accepted Buddhism into its proud history to pretty much take charge of metaphysics. Since I have already spoken of Buddhism, I will include extra-Buddhist China's vision of last things by way of one of my boyhood glimpses into its folk religion.

Directly across from the back gate of our tiny compound in Changshu stood an empty lot that was reserved for public occasions, including funeral rites. For funerals in well-to-do families a life-size paper house was constructed that contained real tables, chairs, and beds (on which I once saw genuine silk coverlets spread). Sheaves of rice straw were propped against the house, and at the appointed moment (while Buddhist priests processed around the house playing flutes and chanting) the makeshift structure was consigned to flames. The obvious point of the ritual (apart from being a display of conspicuous consumption) was to ensure a comfortable hereafter for the deceased. In the last such rite that I witnessed, a papier-mâché replica of a Model-T Ford was parked by the front door, presumably on the assumption that whereas the deceased had only *lived* in this world, he would *really* live in the next.

Now for my promised report on hell. In the West, Christian and Islamic versions of the doctrine hit one in the face—and not without reason, for I am not aware that *eternal* damnation figures outside of them. However, the notion of hell-for-a-spell is widespread. To return for a moment to my boyhood, the large Buddhist temple on whose beautiful grounds our family would often picnic boasted a huge statue of the Buddha that greeted one's eyes when one entered. On the other side of the wall behind that Buddha, however, carved in bas relief a foot deep, was a gigantic panel that depicted the tortures of the damned with a blood-curdling literalness that Hieronymus Bosch would have envied.

To dwell further on damnation here would disrupt this chapter, so I shall close the topic down for now with two staccato points and the promise that I shall return to it in the final chapter of this book. First, psychologists have come to have serious doubts about the efficacy of fear as a deterrent to wrongdoing. Second, in the context of human religiousness as a whole, the doctrine of eternal damnation looks like the exception that proves the rule of a happy ending, rather than a doctrine that retires that rule.

As for the scientific worldview, there is no way that a happy ending can be worked into it. Death is the grim reaper of individual lives, and whether things as a whole will end in a freeze or a fry, with a bang or a whimper (or keep cranking out more insentient matter in an expanding universe) is anybody's guess. Teilhard de Chardin tried heroically to introduce teleology into the universe with his Omega Point, but his vision has passed neither theological nor scientific muster. Theologians want to know where the "fall" and crucifixion are in his scenario, while scientists are downright contemptuous. P. B. Medawar writes of Teilhard's *The Phenomenon of Man,* "Its author can be excused of dishonesty only on the grounds that before deceiving others he has taken great pains to deceive himself."

4. This fourth contrast between the competing worldviews concerns meaning. Having been intentionally created by omnipotent Perfection—or (less anthropomorphically described) flowing as it does from that Perfection "like a fountain ever on," in Plotinus's wording—the traditional world is meaningful throughout. In the scientific worldview, meaning is only skin-deep, "skin" here signifying biological organisms on a single speck in the sidereal universe. As John Avis and William Provine have said, "Our modern understanding of evolution implies that ultimate meaning in life is nonexistent." Steven Weinberg joins them in acknowledging that "the more the universe seems comprehensible, the more it seems pointless."

5. Finally, in the traditional world people feel at home. They *belong* to their world, for they are made of the same spiritually sentient stuff that the world is made of. The *kakemono* (Japanese scroll painting) that hangs in my front hall reminds me daily that "heaven and earth [East Asia's idiom for all that exists] is pervaded with sentience."

We of today can hardly imagine how seamlessly traditional peoples have woven the great world of nature into the spiritual aspects of their lives. To cite a single example, the Pawnee people of Oklahoma pattern their houses after the architecture of nature as they understand it. Still today, often sitting at night on rooftops, children hear from their parents how Evening Star and the Moon created the first girlchild, and Morning Star and the Sun created the first boychild. The Great Chief Star that shines from the direction of the winter wind and never moves is pointed out to them, and until sleep overcomes them, they watch the rest of the stars circle around him. The Great Chief Star reminds tribal chiefs of their responsibility to care for their people. Nothing like this sense of belonging can be derived from the scientific worldview. Albert Camus speaks for its disciples when he says, "If I were a cat I would belong to this world, this world to which I am opposed with the whole of my being." In his ninth decade now, Czeslaw Milosz sees himself as having lived to see the dawning of "the age of homelessness."

Sweetening the Sour Apple

Understandably, there is a tendency to try to soften the stark contours of the modern view and "sweeten the sour apple" (Freud's phrase). Einstein's assertion that "the most beautiful emotion we can experience is the mystical" is regularly quoted in this connection, and with equal regularity it gets updated. As I write these lines, Ursula Goodenough's *The Sacred Depths of Nature* is the current instance. Goodenough admits that her nature has "no Creator, no superordinate meaning of meaning, no purpose other than life's continuance." Still and all, it fills her with feelings of "awe and reverence."

We can be glad that it does, but how much comfort can we draw from that fact when the awe nature awakens in human beings is, like all emotions, no more than a Post-it note, so to speak, affixed to a nature that is unaware of being thus bedecked. Reverence and awe are human sentiments that extend no deeper into nature than human consciousness extends, and in a universe fifteen billion light-years across, that consciousness is a veneer so thin that it approaches a mathematical line. To speak of nature's depths as sacred in themselves, without human beings imputing sanctity to those depths, is to be guilty of the anthropomorphic thinking that John Ruskin dubbed "the pathetic fallacy"—the fallacy of imputing feeling where there is none. Goodenough's "sacredness" is in her eye, the eye of the beholder, and in the eyes of those who share her sensibilities. What is in the depths of nature—its deep structure on which human sentience bobs like a rose petal on the sea—is insentient, quantifiable matter. Professor Goodenough rejects Steven Weinberg's just-cited verdict of meaninglessness, but his is the consistent reading of matter.

Parenthetically, the creativeness of their vocation probably makes it easier for scientists than for most other people to create meaningful lives for themselves, which subjective, existential meaning people like Goodenough project onto the world at large. Around the middle of the twentieth century George Lundberg wrote a much-discussed book titled *Can Science Save Us?* In the verbal storm of yeses and nos that followed, I found my own answer to be this: science *can* save scientists, for the thrill of discovery and the sense that one is onto important things is deeply fulfilling.

A legend that circulated through the corridors of MIT while I was teaching there brings this point home. When Edward Land and his partner were at the decisive point in their discovery of the process that led to the Polaroid camera, they worked around the clock and caught catnaps on their lab tables when they simply had to sleep. At one point Land's partner said he was exhausted and would have to take a break. "Good," Land responded; "we can get our Christmas shopping out of the way." "Ed," his partner said. *"Ed!* It's January 3."

Like all hagiography, the story is probably heavily embroidered, but no one who has been involved in creative work will miss its point. *My* point in relating it is to flag the difference in fulfillment that comes from inventing the Polaroid camera, on the one hand, and buying one, on the other.

How Much Is at Stake.

As I have said, I have tried in this chapter to steer clear of the question of truth and have confined myself to simply contrasting the two worldviews that are contending for the mind of the future. But I have only touched on my reason for adopting this tactic, and the time has come to state it fully.

Where positions are taken for granted, supporting evidence does not enter the picture because the positions seems obviously, self-evidently true. Less commonly recognized than that first point is a second one. In cases where supporting evidence is sought, how seriously it is sought rests on how much depends on the thesis in question being correct. To test the strength of a trouser belt, a hard yank is enough, for the consequence would be minimal if the belt broke. Where lives are at stake, however, the situation changes; thus the strength of parachute ropes must be precisely calibrated.

These two points lock together here as follows. Not being in vogue today, the traditional worldview must now be argued for if it is to remain in the running. Then comes the second point. How seriously we should regard the evidence for or against it depends on how much is at stake. How much is at stake!—I repeat that phrase because it could have served as the title for this second chapter of my book, which argues that the stakes are high. In making that argument, the chapter reenforces the title of the book as a whole, *Why Religion Matters.*

Conclusion

The inherent superiority of the traditional worldview over its scientific alternative has been the running theme of this chapter, but I want to conclude by coming down on it hard. That does not require belaboring the point. A single contrast, followed by the testament of one of the great scientists of the twentieth century, will suffice.

First the contrast. A decade ago a book review in the *Chronicle of Higher Education* opened with this categorical assertion: "If *anything* characterizes 'modernity,' it is the loss of faith in transcendence, in a reality that encompasses and surpasses our quotidian affairs" (italics added). If we set alongside that the testament of a leading English poet of our time, David Gascoyne, we have the point of this chapter in a nutshell: "The underlying theme that has remained constant in almost everything I have written is the intolerable nature of human reality when devoid of all spiritual, metaphysical dimension."

With those two assertions juxtaposed, I offer the verdict of Jacques Monod, the dean of microbiologists when he died a generation ago. It appears as the conclusion of his valedictory book, *Chance and Necessity*.

> No society before ours was ever rent by contradictions so agonizing. In both primitive and classical cultures the animistic tradition [his phrase for what I am calling the traditional worldview] saw knowledge and values stemming from the same source. For the first time in history a civilization is trying to shape itself while clinging desperately to the animistic tradition to justify its values, and at the same time abandoning it as the source of knowledge.
>
> Just as an initial "choice" in the biological evolution of a species can be binding upon its entire future, so the choice of scientific practice, an unconscious choice in the beginning, has launched the evolution of culture on a one-way path: onto a track which nineteenth-century scientism saw leading infallibly upward to an empyrean noon hour for mankind, whereas what we see opening before us today is an abyss of darkness.

The question of whether evidence *requires* that we be rent by this most agonizing contradiction ever, or have fallen into it by a logical mistake, remains for the next chapter.

CHAPTER 3

THE TUNNEL AS SUCH

With questions of truth suspended, the scientific worldview faired poorly in the preceding chapter. Some of its partisans may have gone along with what was being said on the assumption that once truth entered the picture the scientific worldview would regain its stature; if so, the strategy that I announced for that chapter had some effect. Because the traditional worldview is coming from behind, one has now to labor to get it a hearing. Had I started with the question of its truth, buzzwords such as *Copernicus* and *Darwin* would have rushed in and minds would have snapped shut against the past—such at least was my fear.

Ready now to hear what truth has to say on the comparative accuracy of the two worldviews, we find it to be very little—nothing definitive, actually. We generally assume that the findings of science have retired the traditional outlook, but that has been our big mistake, for those findings pertain to the physical universe only—*cosmology,* in my vocabulary—whereas the *metaphysical* question is whether that universe is all that exists. To think that science can speak to that question is like thinking that people floating through space in a huge balloon could use the same flashlight that illumines its interior to see where the balloon is situated in space. Or (to change the analogy), it is as if, on hearing that the boys on such-and-such block have eyes only for Suzzie, the hearer were to conclude that Suzzie was the only girl on that block.

If science cannot tell us what (if anything) is outside our universe, what can? Nothing *definitively*, but it would be foolish not to draw on every resource available. Inclusively, things are neither as science says they are nor as religion says they are. They are as science, and religion, and philosophy, and art, and common sense, and our deepest intuitions, and our practiced imaginations say they are. What all of these complementing resources—with the exception of modern science, which works with a limited viewfinder (see Chapter Twelve)—have said about the Big Picture throughout human history has shaken down into a single, wondrously clear and inspiring worldview. This worldview, which I consider the winnowed wisdom of the human race, is found distilled in the world's great, enduring religions.

I myself consider this convergent report to be the best measure of truth about the whole of things that we have, but I cannot prove that, so I will say only one more thing about truth before turning to this chapter's main concern. The pragmatic theory of truth defines it as *what works.* I am not fond of that theory, but as long as we do not allow it the last word, it gives us things to think about, and here its interesting deliverance concerns the placebo effect. Physicians have found no remedy to be as universally effective as the placebo. Psychologically that translates into *If you think something will help, it does help;* or, more generally, *If you think affirmatively, your immune system responds affirmatively.*

If that is the psychosomatic truth of the matter, its metaphysical extension is that an affirmative attitude toward life pays off. It seems safe to say that the five-point comparison of the traditional and scientific worldviews in the preceding chapter shows the former to be more conducive to a positive lifestance than is the latter, and the following three almost randomly chosen facts speak to consequences of such a stance:

- A graduate student in psychology at New York University conducted an experiment on students who were taking a course on business law. Dividing them into an experimental and a control group, he had both groups gaze in separate rooms at blank screens for a minute before they entered the classroom for each class session. For four microseconds—too short a time to be consciously perceived—a tachistoscopic image was flashed on each screen. For the experimental group it read, "Mommie and I Are One"; for the control group, "People Are Walking." When this summer course ended, it was found that students in the experimental group scored almost a full letter grade higher in the course.
- Robert Rosenthal's famous "Pygmalion in the classroom" experiments show that when teachers raise their expectations of certain students, those students pick this up, and the improved self-image that the elevated expectations accord them causes them to perform better than they previously had.
- A 1999 Duke University study shows that regular churchgoers are 28 percent less likely to die in a given seven-year period than non-churchgoers. Many studies of this sort have been conducted. All that I have seen are in line with this one.

Pragmatic considerations such as these do, as I say, give us things to think about, but I want to shut down on them fast. The problem is not that if one tries to make much of them, mountains of contingencies pour in to blur the claims being made. More central is the fact that using the consequences of religious beliefs to support the beliefs themselves will not work because placebos are effective only when they are not known to be such. If the traditional worldview is to have beneficial effects, it will be because it is believed to be true, and one cannot argue oneself into believing that X is true because it pays dividends. The Duke study illustrates a second danger. When the consequences of belief are worldly goods, such as health, fixing on them turns religion into a service station for self-gratification and churches into health clubs. This is the opposite of authentic religion's role, which is to de-center the ego, not pander to its worldly desires.

With these epistemological points entered, I proceed now to the tunnel that science-decked-out-as-metaphysics has led us into. The preceding chapter positioned the tunnel in the terrain it runs through, this chapter will consider it in its entirety, and the four chapters that follow will describe its sides. All five of these tunnel-related chapters will begin with a book I have selected to serve as the chapter's flagship, and I will follow their respective wakes so faithfully that the reader could do worse than to think of this first half of my book as a mosaic configured from the books I use to introduce its respective chapters. I proceed to the book chosen for this chapter.

The Flagship Book

William Gass's 1995 *The Tunnel* was a strong contender for the role. It's title is the operative metaphor for my entire book, and his book is oppressively gloomy in the way tunnels always are. Its protagonist is as repulsively lonely a character as fiction affords. A middle-aged professor of history at a midwestern university, he takes to going down into the cellar of his big middle-class house to escape from his unloved, undesired, and unloving wife. He starts tunneling down through the floor and out beyond its foundations, lying on his fat belly and squirming past trowelfuls of clay and dirt and dust on his way out. He is trying to escape from his life and from our times, which his horrible home symbolizes.

This storyline fits this third chapter of my own book so snugly that, if it were not for two clear differences, I could have been tempted into thinking that the secret reason for Gass's writing his novel was to pipe my book on stage. The first of these differences is that Gass's is an aristocratic book, written for the literary elite, whereas mine is as plebeian as I can render its not always simple arguments. And his book is postmodern—indeed, aggressively so. Heralded by some as *the* postmodern novel, it is intentionally multivalent and opaque to allow for alternative readings at every turn. My aim is the opposite—to be as clear and direct as the subject matter allows.

With *The Tunnel* disqualified, I reached for T. S. Eliot's *The Waste Land*—from a five-foot shelf of candidates, I hasten to say—because an important sign that we are *in* a tunnel is the way the twentieth century replaced utopias with dystopias. The century in which politicians preyed on hope as never before—promising "the war to end all wars," "the war to make the world safe for democracy," "the century of the common man," "the four freedoms," "one world," "the great society," and "the new world order"—saw utopian writing come to a dead halt. Nietzsche foresaw the twentieth century as clearly as anyone, and we have his verdict: "The very best is not to be born; the next best is to die soon." For Yeats, "the best lack all conviction, while the worst are full of passionate intensity." Kazantzakis concluded that "hope is a rotten-thighed whore," and even Bergson (who moved Darwin's onward-and-upward biology into philosophy) was driven in the end to the view that human beings were being crushed by the immense progress they had made. I have never managed to think of Sartre as profound, but he was a shrewd phenomenologist, and on the existential level where he worked, he concluded that "we must learn to live without hope." Motion picture titles take up the refrain: *I Have Seen the Future and It Doesn't Work.*

Given pronouncements of this sort, I could have chosen any number of books to steer this chapter by: Arthur Koestler's *Darkness at Noon,* Samuel Beckett's *Waiting for Godot,* W. H. Auden's *The Age of Anxiety,* Aldous Huxley's *Brave New World,* George Orwell's *1984,* C. S. Lewis's *The Abolition of Man,* Herbert Marcuse's *One Dimensional Man,* or countless others. Instead, I choose *The Waste Land,* because its title tells the story by itself and (as a supporting reason) because it pairs logically with its sequel, *The Hollow Men,* to give us a brace of books (in this case poems) to pace this chapter. The familiar opening stanzas of the poems read like epitaphs for the twentieth century. From *The Waste Land:*

April is the cruelest month, breeding
Lilacs out of the dead land. . . .
What . . . branches grow
Out of this stony rubbish? Son of man,
You cannot say, or guess, for you know only
A heap of broken images, where the sun beats,
And the dead tree gives no shelter, the cricket no relief,
And the dry stone no sound of water. . . .

The Hollow Men describes the denizens of this waste land:

We are the hollow men
We are the stuffed men
Leaning together
Headpieces filled with straw. Alas!
Our dried voices, when
We whisper together
Are quiet and meaningless
As wind in dry grass
Or rat's feet over broken glass
In our dry cellar.

The balance of this chapter expands Eliot's lines in my own way.

The Tunnel in Question

I remind the reader that (as I said in Chapter Two) the tunnel metaphor in this book applies not to our times as a whole but to the worldview into which we have unwittingly wandered. This presents a difficulty, for (as that chapter went on to say) worldviews tend to pass unnoticed. In traditional times this did not much matter, for then the views were life-enhancing. The important buildings were temples; statues were of gods and saints; legends, songs, and dances wore the cast of morality plays; and holidays lived up to being holy days. Reminders of the sacred were everywhere, strewn about almost carelessly, we might say. Marco Pallis reported that in the traditional Tibet that he knew the entire landscape seemed to be suffused by the message of the Buddha's teachings. "It came to one with the air one breathed. Birds seemed to sing of it; mountain streams hummed its refrain as they bubbled across the stones. A holy perfume seemed to rise from every flower, at once a reminder and a pointer to what still needed doing. There were times when a man might have been forgiven for supposing himself already in the Pure Land."

In times like those, explicit references to the sacred were hardly necessary, but those times are long gone. Today we do not live under a sacred canopy; it is marketing that forms the backdrop of our culture. The message that advertising dins into our conscious and unconscious minds is that fulfillment derives from the things we possess. Because this is not true, the message serves us badly, so we need to be aware of the worldview that sponsors it. The line that runs from a materialistic worldview to a materialistic philosophy of life is not a straight one, but it exists all the same. Unfortunately, philosophers, who formerly made it their job to monitor worldviews and their consequences, have now disclaimed that responsibility. As early as the middle of the nineteenth century, Jacques Maritain was warning that "a weakening of the metaphysical spirit is an incalculable damage for the general order of intelligence and human affairs," but philosophers were in no mood to listen and opted for the post-Nietzschean deconstruction of metaphysics. "We live in an age when the very possibility of metaphysics is hardly admitted without a struggle," R. G. Collingwood noted, and Iris Murdoch seconded his observation. "Modern philosophy is profoundly anti-metaphysical in spirit," she remarked over her shoulder as she poured most of her energy into writing novels.

This rejection by philosophers of what historically had been their main contribution to civilization is curious, but it does not lack for an explanation—two related explanations, in fact. First, mistaking cosmology for metaphysics—the mistake that modernity as a whole has made—contemporary philosophers tend to assume that scientists are in a better position to see the whole of things than they themselves are. The following remark by John Searle brings this out explicitly: "Most professionals in philosophy, psychology, artificial intelligence, neurobiology, and cognitive science accept some version of materialism because they believe that it is the only philosophy that is consistent with our contemporary scientific worldview."

The companion reason for metaphysics' demise is postmodernism, for (as Chapter One noted) it emerged in good part to do the project in. Assuming without argument that worldviews necessarily oppress, and overlooking the fact that even if that were the case, they cannot be excised from human knowing, philosophers have tried to manufacture a metaphysics-less world, an oxymoron if there ever was one. Several years back the University of Chicago alumni magazine featured Richard Rorty on the cover of one of its issues announcing that "there is no Big Picture."

As that is not the case, I continue with the task that constitutes this book's critical side—that of holding modernity's Big Picture up to withering inspection. (Modernity and postmodernity cannot be sealed from each other. The context must determine whether I am using the word *modernity* to refer to the period that preceded postmodernity, as I do in Chapter One, or to include postmodernity, as I use the word here.)

A Disqualified Universe

Lewis Mumford's memorable characterization of the scientific worldview as "disqualified" rides a play on the word. That worldview is disqualified in the straightforward sense of being "out of the running" as a human home, and what disqualifies it for that role is the way it strips the objective world of its qualities and leaves it "dis-qualified" in that awkward but altogether appropriate sense of the word .

We commonly assume that science can at least handle the corporeal world that our physical senses register, but strictly speaking that is not the case, for we experience the corporeal world decked with sounds, smells, and colors, whereas science gives us only the quantifiable underpinnings of those sensations. "Secondary qualities"—the colors we see and birdsongs we hear—do not turn up in science's textbooks. From its standpoint, human beings (and perhaps other animals) paint those qualities onto the world, so to speak.

And if secondary qualities have no place in the objective, transhuman world, much less do "tertiary qualities"—which is to say, values. Hopes and fears, pleasures and pains, successes and disappointments—the sum total of the lives that we experience directly—are for science epiphenomenal only, the foam on the beer, which requires beer (matter) to exist but not vice versa. "Only connect," E. M. Forster counseled as his valediction, but what is there to connect to when what is distinctively human about us is only skin deep in the objective nature of things?

Forster's succinct counsel strikes an important note that is worth pursuing. Thanks to the marvels of microphotography we can now see single nerve cells, and what catches the eye is their dendrites, waving in the air like the tendrils of sea anemones in the hope—so it appears—of touching the dendrites of another cell. When two dendrites do touch, they lock arms and, as a result, their cells stand a better chance of braving life's perils. It is religion in embryo, for *religio* in Latin means "to rebind," and bonding and rebinding are what religion is all about.

Traditional societies find bondedness built into the fabric of things, and they use their religions to keep the world from unraveling. Religions show people bonded to the ultimate Source of things by their very lineage, for if the Ultimate did not literally parent them (as Izanagi and Izanami did in Japan's creation myth), it "begot" them in substance through creation or emanation. And because human beings have derived from bonding, it becomes incumbent on them to bond with others. "Be ye members one of another," St. Paul counseled. Confucius's version reads, "Within the four seas all men are brothers."

Nature is included in the picture as well. The title of Carolyn Merchant's *The Death of Nature* reminds us that nature was not always thought of as dead. The earth was seen as alive and considered to be a beneficent, receptive, nurturing female. The Roman compiler Pliny warned against mining the depths of Mother Earth, speculating that earthquakes were an expression of her indignation at being violated in this manner. In much the same way, Native Americans objected to European ways of treating the Earth Mother. In the words of the Smohalla of the Columbia Basin tribes:

You ask me to plow the ground! Shall I take a knife and tear my mother's breast?
You ask me to dig for stone! Shall I dig under her skin for her bones?
You ask me to cut grass and make hay and sell it! How dare I cut off my mother's hair?

The reference to Confucius's "four seas" aphorism in an earlier paragraph triggers a recollection that relates to it. Several years ago my wife, Kendra, took a young grandson to the neighborhood playground where they found two children already on the swings and slides—a girl about eight and a younger boy, presumably her brother. With the briefest of preliminaries the girl asked Kendra, "What are we?" Kendra squinted a bit and answered, "Chinese?" "No." "Vietnamese?" "No," with a touch of irritation entering. When Kendra ventured a third mistaken possibility, the irritation erupted. "*No!* What *are* we?" At that point Kendra (thinking that if she knew the answer she might better understand the question) said, "I give up. What are you?" "We are brother and sister," the girl replied, "and so we love each other. And our grandmother tells us that if we love her, when we become grandparents our grandchildren will love us."

In our jaded, individualistic society, it may take a child—perhaps one with lingering ties to a traditional civilization—to come right to the mark. Not *who* are we, which points toward differences, but *what* are we; what is our basic essence? And the youngster's answer was equally on the mark. Our essence is relationship—we are brother and sister—and the foundation of that essence is love.

What happens when the sense of bondedness to the Ultimate erodes and religious directives to bond are discounted? Already a century ago, W. B. Yeats was warning that things were falling apart, that the center did not hold. Gertrude Stein followed him by noting that "in the twentieth century nothing is in agreement with anything else." Ezra Pound saw man as "hurling himself at indomitable chaos," and the most durable line from the play *Green Pastures* has been, "Everything that's tied down is coming loose." It is not surprising, therefore, that when in her last interview Rebecca West was asked to name the dominant mood of her time, she replied, "A desperate search for a pattern." The search is desperate because it seems futile to look for a pattern when reality has become, in Roland Barthes's vivid image, kaleidoscopic. With every tick of the clock the pieces of experience come down in new array.

I am in danger of overstating my case. The direct cause of the dislocations that the twentieth century experienced was technology. Its automobiles weakened extended families and undermined communities, and its radios and televisions relieved people of the effort required to get together to create their own entertainments. Changes such as these, more than the shift in worldviews, are primarily responsible for creating the most individualistic society history has known—and I promised to stick to worldviews. So I will revert to the minimalist claim for this book that I registered earlier: whatever else has gone on and is going on, we would be better off with a worldview that shows us deeply connected to the final nature of things. It is cruel to heap psychic pain on physical pain. It does not condone slavery in the slightest to ask rhetorically if the spirituals that slaves composed and sang in the cotton fields—"Swing Low, Sweet Chariot," "Go Down, Moses," "Take My Hand, Gentle Lord"—did not help them to endure their intolerable hardships. The same can be asked regarding the Holocaust victims who had God on their lips when they died.

And while I am disclaiming, I want again to exonerate scientists as a professional group from having gotten us into our tunnel. When modern science gained momentum and took off, its payoffs (noetic as much as technological) were so impressive that it looked as if they might bring heaven to earth, as Carl Becker's minor classic, *The Heavenly City of the Eighteenth Century Philosophers,* put the matter. We should never forget that those inflated expectations of science were powered by *everyone's* hopes. Scientists did not push us into the tunnel; it is more accurate to say that we the people rushed headlong for it and all but dragged scientists with us. Malice nowhere entered. Only consequences unforeseen, and the small but fateful slip in logic that has been noted.

With scientists exonerated from getting us into our tunnel—they bear some responsibility for *keeping* us in it, but that is for the next chapter—I return to the feel of the tunnel proper. As Chapter Two pointed out, in the great outdoors people felt at home. It was made of the same stuff—sentience, values, meanings, and purposes—that we are made of. It was under able management. And because consciousness (not matter) was foundational, bodily death did not have the final say. In stark contrast to this, there is no way that a tunnel can feel like a home. In the scientific worldview everything derives from and depends on inert matter, and apart from organic life, purposelessness reigns.

We see the contrast clearly in the fate that has befallen Aristotle's doctrine of the four causes: material, efficient, formal, and final. The *material* cause of my computer is (in part) its silicon chips; its *efficient* cause was the labor of the people who assembled it; its *formal* cause was the design that guided their actions; and its *final* cause was their objective in producing such machines—to help me write this book, in the case in hand. Science retains the first two of the four but restricts the other two to organisms—that razor-thin veneer on the world of dead matter. The hostility of hard-nosed scientists to the notion of "intelligent design" is part and parcel of their denial that formal causes exist anywhere except in human minds;

and though *teleonomy* is useful in describing the purposive behavior of organisms, apart from that special case *teleology* is out. "The cornerstone of scientific method," Jacques Monod asserted, "is the *systematic* denial that 'true' knowledge can be got at by interpreting phenomena in terms of final causes—that is to say, of 'purpose.'"

With purposelessness (and its synonym *chance*) in the driver's seat, blind whirl is king. This leaves us "strangers and afraid in a world [we] never made," to quote A. E. Housman's "The Shropshire Lad." Albert Camus found the world "absurd." Samuel Beckett's spokesmen wait out their lives for a what-they-know-not that never arrives. Franz Kafka concluded that "the world order is a lie." Two allegories frame Western civilization like majestic bookends—Plato's allegory of the cave at its start, and at its end-thus-far Nietzsche's madman charging through the streets announcing that God is dead. There could hardly be two more telling characterizations of tradition's great outdoors and modernity's tunnel.

It could have been expected that the inhospitality of the scientific world to humanity's deepest concerns would generate revolts. Romanticism and existentialism were the chief of these. William Blake called for "the rise of soul against intellect" to save us from "Single vision and Newton's Sleep!" but Matthew Arnold's lament over "faith's long receding roar" a century later amounted to an admission that Blake's cause was not succeeding. As for existentialism, it held out resolutely for human freedom in the face of science's seemingly deterministic world, but neither it nor romanticism was able to stem the scientific tide because they had no alternative worldview in which to anchor the rights of the human that they so nobly championed. Belonging to the extrascientific side of culture, they carried no weight in the scientific side, which settled its questions within itself, marshaling evidence powerful enough to flatten cities and bore holes in steel with drills of light.

Conclusion

An interview in the issue of the *New Yorker* that is on the stands as I conclude this chapter has Albert Gore pointing to "a kind of psychic pain at the very root of the modern mind." It is poets, though, not politicians, who deserve the last word, so I will invoke two of them to round off this chapter.

Bertolt Brecht is best remembered for his plays, but critics consider his poems more profound. The relevant one here is titled "To Those Born Later":

Truly I live in dark times!
The innocent word is folly.
An unlined forehead
Suggests insensitivity.
The man who laughs
Just hasn't heard
The terrible news.

Stephen Dunn is less well known, and the style of his poetry is as different from Brecht's as one could imagine. In its own way, however, it too works to pull the contents of this chapter together, so I shall quote "At the Smithville Methodist Church" in full:

It was supposed to be Arts and Crafts for a week,
but when she came home
with the "Jesus Saves" button, we knew what art
was up, what ancient craft.

She liked her little friends. She liked the songs
they sang when they weren't
twisting and folding paper into dolls.
What could be so bad?

Jesus had been a good man, and putting faith
in good men was what
we had to do to stay this side of cynicism,
that other sadness.

O.K., we said, one week. But when she came home
singing "Jesus loves me,
the Bible tells me so," it was time to talk.
Could we say Jesus

doesn't love you? Could I tell her the Bible
is a great book certain people use
to make you feel bad? We sent her back
without a word.

It had been so long since we believed, so long
since we needed Jesus
as our nemesis and friend, that we thought he was
sufficiently dead,

that our children would think of him like Lincoln
or Thomas Jefferson.
Soon it became clear to us: you can't teach disbelief
to a child,

only wonderful stories, and we hadn't a story
nearly as good.
On parents' night there were the Arts and Crafts
all spread out

like appetizers. Then we took our seats
in the church
and the children sang a song about the Ark,
and Hallelujah

and one in which they had to jump up and down
for Jesus,
I can't remember ever feeling so uncertain
about what's comic, what's serious.

Evolution is magical but devoid of heroes.
You can't say to your child
"Evolution loves you." The story stinks
of extinction and nothing

exciting happens for centuries. I didn't have
a wonderful story for my child
and she was beaming. All the way home in the car
she sang the songs,

occasionally standing up for Jesus.
There was nothing to do
but drive, ride it out, sing along
in silence.

"We hadn't a story nearly as good." We need not restrict ourselves to the Jesus story in this, for its counterparts turn up in every tribe and civilization. Jews have their Passover story of a miraculous escape from Egypt. In the Bhagavad Gita, Arjuna, on the eve of a horrendous battle, wrings the meaning of life and death from Krishna, disguised as his charioteer. The Jataka Tales have Siddhartha Gautama in his incarnation as a rabbit throwing himself on a fire to save luckless hunters from starving. The list has no end.

CHAPTER 4

The Tunnel's Floor: Scientism

With the foregoing account of the tunnel in place, I proceed now to describe its four sides, beginning with the floor—scientism—which supports the other three sides. Only four letters, "tism," separate scientism from science, but that small slip twixt the cup and the lip is the cause of all our current problems relating to worldview and the human spirit. Science is on balance good, whereas nothing good can be said for scientism.

Everything depends on definitions here, for this chapter will fall apart if the distinction between *science* and *scientism* is allowed to slip from view. To get those definitions right requires cutting through the swarm of thoughts, images, sentiments, and vested interests that circle the word *science* today to arrive at the only definition of the word that I take to be incontrovertible—namely, that science is what has changed our world. Accompanied by technology (its spin-off), modern science is what divides modern from traditional societies and civilizations. Its content is the body of facts about the natural world that the scientific method has brought to light, the crux of that method being the controlled experiment with its capacity to winnow true from false hypotheses about the empirical world.

Scientism adds to science two corollaries: first, that the scientific method is, if not the *only* reliable method of getting at truth, then at least the *most* reliable method; and second, that the things science deals with—material entities—are the most fundamental things that exist. These two corollaries are seldom voiced, for once they are brought to attention it is not difficult to see that they are arbitrary. Unsupported by facts, they are at best philosophical assumptions and at worst merely opinions. This book will be peppered with instances of scientism, and one of Freud's assertions can head the parade: "Our science is not illusion, but an illusion it would be to suppose that what science cannot give us we can get elsewhere." Our ethos teeters precariously on sandy foundations such as this.

So important and undernoticed is this fact that I shall devote another paragraph to stating it more concretely. For the knowledge class in our industrialized Western civilization, it has come to seem self-evident that the scientific account of the world gives us its full story and that the supposed transcendent realities of which religions speak are at best doubtful. If in any way our hopes, dreams, intuitions, glimpses of transcendence, intimations of immortality, and mystical experiences break step with this view of things, they are overshadowed by the scientific account. Yet history is a graveyard for outlooks that were once taken for granted. Today's common sense becomes tomorrow's laughingstock; time makes ancient truth uncouth. Einstein defined common sense as what we are taught by the age of six, or perhaps fourteen in the case of complex ideas. Wisdom begins with the recognition that our presuppositions are options that can be examined and replaced if found wanting.

The Flagship Book

My flagship book for this chapter is Bryan Appleyard's *Understanding the Present: Science and the Soul of Modern Man.* I will compress its thesis into a story, the details of which are mine, but whose plot is his.

Imagine a missionary to Africa. Conversion is slow going until a child comes down with an infectious disease. The tribal doctors are summoned, but to no avail; life is draining from the hapless infant. At that point the missionary remembers that at the last minute she slipped some penicillin into her travel bags. She administers it and the child recovers. With that single act, says Appleyard, it is all over for the tribal culture. Elijah (modern science) has met the prophets of Baal, and Elijah has triumphed.

If only that tribe could have reasoned as follows, Appleyard continues; if only they could have said to themselves, This foreigner obviously knows things about our bodies that we do not know, and we should be very grateful to her for coming all this distance to share her knowledge with us. But as her medicine appears to tell us nothing about who we are, where we came from, why we are here, what we should be doing while we are here (if anything), and what happens to us when we die, there seems to be no reason why we cannot accept her medicine gratefully while continuing to honor the great orienting myths that our ancestors have handed down to us and that give meaning and motivation to our lives.

If only those tribal leaders had the wit to reason in that fashion, Appleyard concludes, there would be no problem. But they do not have that wit, and neither do we.

From that fictionalized condensation of Appleyard's book, I proceed to develop its thesis in my own way, beginning with the reception his book received.

Before I had laid hands on Appleyard's book, I attended a conference at the University of Notre Dame. Finding myself at breakfast one morning with the noted British scientist Arthur Peacocke, I asked him about the book, for it had first appeared in England and I thought Peacocke might have gotten the jump on me in reading it. He said that he had not read it but had heard that it was an anti-science book.

Click! Scientism. Scientism, because when I got to the book it turned out not to be against science at all, not science distinct from scientism. But because it spells out with unusual force and clarity what social critics have been saying for some time now—namely, that we have turned science into a sacred cow and are suffering the consequences idolatry invariably exacts, it is a sitting duck to be taken as an attack on the scientific enterprise. Not by all scientists. It is not a digression to say (before I continue with Appleyard) that not all scientists idolize their profession. The spring 1999 issue of the *American Scholar* that crosses my desk on the day that I write this page bears this out forcefully. Its review of *Of Flies, Mice, and Men* sees its author, the French microbiologist François Jacob, as having written his book "to renounce much of the epistemological privilege of science, for as [he] points out with surprising and even extreme determination, the myths, misconceptions, and misuses of science can be insidious. They infiltrate out language and beliefs even as we try to expel them."

I could hardly ask for a stronger ally in this chapter than biologist Jacob, and with his support I return to Brian Appleyard.

When *Understanding the Present* was published, responses to it polarized immediately. The *Times Literary Review* saw the book's author as voicing truths that needed to be spoken, whereas England's leading scientific journal, *Nature,* branded it "dangerous."

When reviews began to appear on this side of the Atlantic, the *New York Review of Books* chose a science writer, Timothy Ferris, to do the job. Ferris gives us his opinion of the book in his closing paragraph. "Its real target," he writes, "would appear to be not science but scientism, the belief that science provides not *a* path to truth, but the *only* path." So far, fair enough—but then Ferris tells us that

> scientism flourished briefly in the nineteenth century, when a few thinkers, impressed by such triumphs as Newtonian dynamics and the second law of thermodynamics, permitted themselves to imagine that science might soon be able to predict everything, and we ought to be able to muster the sophistication to recognize such claims as hyperbolic. Scientism today is advocated by only a tiny minority of scientists.

Those of us who stand outside the science camp can only read such words with astonishment. "Scientism flourished *briefly* when *a few* thinkers permitted themselves to imagine that science might soon be able to predict everything"? "Scientism today is advocated by only *a tiny minority* of scientists"? Ferris's assertions dismiss *the* metaphysical problem of our time by definitional fiat, for if you define scientism as the belief "that science might soon be able to predict everything," then of course too few people believe *that* for it to constitute a problem.

Tracking Scientism

A discussion I was party to recently comes to mind. Historians of religion were asking themselves why the passion for justice surfaces more strongly in the Hebrew scriptures than in others, and when someone came up with the answer it seemed obvious to us all. No other sacred text was assembled by a people who had suffered as much *injustice* as the Jews had, and this made them privy from the inside to the pain injustice occasions. It is extravagant to compare the damage that scientism wreaks to the suffering of the Jews, but the underlying principle is the same in both cases. Only discerning victims of scientism (and sensitive scientists like François Jacob whom I quoted several paragraphs back) can comprehend the magnitude of its oppressive force and the problems it creates. For it takes an eye like the one Michel Foucault trained on prisons, mental institutions, and hospitals (which eye I am striving for in this book) to detect the power plays that the micro-practices of scientism exert in contemporary life.

Another procedural point must be entered, for it too is often overlooked. What is and is not seen to be scientism is itself metaphysically controlled, for if one believes that the scientific worldview is true, the two appendages to it that turn it into scientism are not seen to be opinions. (I remind the reader that the appendages are, first, that science is our best window onto the world and, second, that matter is the foundation of everything that exists.) They present themselves as facts. That they are not provable does not count against them, because they are taken to be self-evident—as plainly so as the proverbial hand before one's face.

This poses *the* major problem for this book, because what is taken to be self-evident depends on one's worldview, and disputes among worldviews are (as the preceding chapter established) unresolvable. Today's science-backed self-evidence is a fact of contemporary life that must be lived with. It is like wind in one's face on a long journey: to be faced without allowing it to divert one from one's intended course. During the McCarthy era it was said that Joe McCarthy found Communists under every bed, and those who are on the science side in this debate will see me as doing the same with scientism—or as finding under stones the sermons I have already put there, as Oscar Wilde charged Wordsworth with doing. There being (from their point of view) no problem, they will see this entire book as an exercise in paranoia. Because the difference comes down to one of perception, I will plow ahead in the face of that charge, taking heart from the way Peter Drucker perceived his vocation.

As the dean of management consultants in their founding generation, Drucker received every honor that his field had to confer.

When he retired, he was asked in an interview if there was anything professionally that he would have liked to have had happen that had not happened. Drucker answered that actually there was. He kept replaying in his mind a scenario that in real life had never transpired. In it he was seated with the CEO of a company in the wrap-up session of a two-week consultation. Having looked together into every aspect of the company's operations they could think of, the two had become friends and grown used to speaking frankly to each other, so at one point the CEO leans back in his chair and says, "Peter, you haven't told me a thing I didn't already know."

"Because," Drucker added, "that's invariably the case. I never tell my clients anything they don't already know. My job is to make them see that what they have been dismissing as incidental evidence is actually crucial evidence." That is what I see myself doing with respect to scientism in this book.

Having referred to the *New York Review of Books* regarding its handling of Appleyard's book, I will turn to it again for my next example of scientism, for that journal serves as something of a house organ for the elite reading public in America.

John Polkinghorne is a ranking British scientist who at the age of fifty became an Anglican clergyman. The *New York Review of Books* never reviews theological books; but presumably because Polkinghorne is also a distinguished scientist, it made an exception in his case. To review his book, the *NYRB* reached for a world-class scientist, Freeman Dyson. *Click!* A scientist to review a book on theology? To see what that choice bespeaks, we need only turn the table and try to imagine the editors of the *NYRB* reaching for a theologian to review a book on science. The standard justification for this asymmetry is that science is a technical subject whereas theology is not, but now hear this. Several years back at a conference at Notre Dame University I heard a leading Thomist say in an aside to the paper he was delivering, "There may be—there just *may* be—twelve scholars alive today who understand St. Thomas, and I am not one of them."

We turn now to what Dyson said about Polkinghorne's book. After commending its author for his contributions to science and for historical sections of the book under review, Dyson turned to his theology, which like all theology, he said, suffers from being about words only, whereas science is about things. *Click* and *double-click!* As a self-appointed watchdog on scientism, I took pen in hand and challenged that claim in a letter to the *NYRB* that began as follows:

> It is symptomatic of the unlevel playing field on which science and religion contend today that a scientist with no theological credentials (Freeman Dyson in the *New York Review,* May 28, 1998) feels comfortable in concluding that the theology of a fellow scientist (John Polkinghorne) is, like all theology, about words and not, as is the case with science, about things. This flies in the face of the fact that most theology takes God to be the only completely real "thing" there is, all else being like shadows in Plato's cave. Muslims in their testament of faith sometimes transpose "There is no God but the God" to read, "There is no Reality but *the* Reality," the two assertions being identical.

The rest of my letter is irrelevant here, but I do want to quote the first sentences of Freeman Dyson's reply as indicative of the graciousness of the man. "I am grateful to Huston Smith for correcting my mistakes," he wrote. "I have, as he says, no theological credentials. I have learned a lot from his letter." Dyson may have no theological credentials, but he is certainly a gentleman.

In a chapter that has to struggle at every turn not to sound peevish and aggrieved, whimsy helps, so I will mention the occasion on which I found scientism aimed most pointedly (though disarmingly) at me. (I told the story in my *Forgotten Truth,* but it bears repeating here.)

Not surprisingly, the incident took place at MIT, where I taught for fifteen years. I was lunching at the faculty club and found myself seated next to a scientist. As often happened in such circumstances, the conversation turned to the differences between science and the humanities. We were getting nowhere when suddenly my conversational partner interrupted what I was saying with the authority of a man who had discovered Truth. "I have it!" he exclaimed. "The difference between us is that I count and you don't." Touché! Numbers being the language of science, he had compressed the difference between C. P. Snow's "two cultures" into a double entendre.

The tone in which his discovery was delivered—playful, but with a point—helped, as it did on another MIT occasion. When I asked a scientist how he and his colleagues regarded us humanists, he answered affably, "We don't even bother to ignore you guys." Despite the levity in these accounts, the very telling of them opens me to the charge of sour grapes, so to those who will say that I am embittered I will say that they are quite wrong. Our scientific age has, if anything, treated me personally above my due. My concern is with scientism's effect on our time, our collective mindset—the fact that (to go back to Appleyard) it is "spiritually corrosive, and, having wrestled religion off the mat, burns away ancient authorities and traditions." The chief way it does this, Appleyard continues, "is by separating our values from our knowledge of the world." Timothy Ferris dismisses this charge as "extravagant and empty," and here again we can only be astonished at how blind those inside the scientific worldview are to the scientism that others find riddling modernism throughout. For, science writer that he is, there is no way Ferris could have been unaware that Jacques Monod drew a gloomier conclusion from our having separated values from knowledge than Appleyard does. Think back to the key assertions by Monod that I used to conclude Chapter Two: "No society before ours was ever rent by contradictions so agonizing. . . . What we see before us is an abyss of darkness."

Thus far this chapter has proceeded largely in the wake of Appleyard's book. I want soon to strike out on my own, but not before adding Appleyard's most emphatic charge, which is that "science has shown itself unable to coexist with anything." Science swallows the world, or at least more than its share of it. Appleyard does not mention Spinoza in this connection, but I find in Spinoza's *conatus* the reason for Appleyard's charge.

Spinoza's Conatus

Spinoza wrote in Latin, and the Latin word *conatus* translates into English as "will." Every organism, Spinoza argued, has within it a will to expand its turf until it bumps into something that stops it, saying to it, in effect, *Stay out; that's my turf you're trespassing on.* Spinoza did not extend his point to institutions, but it applies equally to them, and I find in this the explanation for why science has not yet learned the art of coexistence. Most scientists as individuals have mastered that art, but when they gather in institutions—the appropriately named American Association for the Advancement of Science, the *Scientific American,* and the like—collegiality takes over and one feels like a traitor if one does not pitch in to advance one's profession's prestige, power, and pay. I have a friend who is an airline pilot who flies jumbo jets. At the moment, his union is threatening to strike for a pay increase. He personally thinks that pilots are already overpaid and is free to say that and vote against the strike in union meetings. But if the motion to strike carries, he will be out there on the picket line, waiving his striker's placard. It is this—group dynamics, if you will—not the arrogance of individuals, that explains why science, which now holds the cards, "has shown itself unable to exist with anything." There is no institution today that has the power to say to science, *Stand back; that's my turf you're poaching on.*

I can remember the exact moment when this important fact broke over me like an epiphany. It was a decade or so ago and I was leading an all-day seminar on scientism in Ojai, California. As the day progressed, I found myself becoming increasingly aware of a relatively young man in the audience who seemed to be taking in every word I said without saying a word himself. True to form, when the seminar ended in the late afternoon, he held back until others had tendered their goodbyes, whereupon he asked if I would like to join him for a walk. The weather was beautiful and we had been sitting all day, but it was primarily because I had grown curious about the man that I readily accepted his invitation.

He turned out to be a professor at the University of Minnesota whose job was teaching science to nonscientists. Word of my seminar had crossed his desk, and being invested in the topic, he had flown out for the weekend. "You handled the subject well today," he said, after we had put preliminaries behind us, "but there's one thing about scientism that you still don't see. Huston, science *is* scientism."

At first that sounded odd to me, for I had devoted the entire day to distinguishing the two as sharply as I could. Quickly, though, I saw his point. I had been speaking *de jure* and completely omitting the *de facto* side of the story. In principle it is easy to distinguish science from scientism. All the while, in practice—in the way scientism works itself out in our society—the separation is impossible. Science's *conatus* inevitably enters the picture, as it does in every institution. The American Medical Association is an obvious example, but the signs are everywhere.

Jürgen Habermas, a philosopher of the Frankfurt School, coined a useful phrase for the way money, power, and technology have adversely affected the conditions of communication in ordinary, face-to-face life. He charged them with "colonizing the life world." A neo-Marxist himself, he had no particular interest in religion, but the concerns of this book prompt me to add scientism to his list of imperialists. One of the subtlest, most subversive ways it proceeds is by paying lip service to religion while demoting it. An instance of this is Stephen Jay Gould's book *Rocks of Ages,* which I will approach by way of a flashback to Lyndon Johnson. It is reported that when a certain congressman did something President Johnson considered reprehensible, Johnson called him into his office and said, "First I'm going to preach you a nice little sermon on how that's not the way to behave. And then I'm going to ruin you."

My nice little sermon to Professor Gould is, "Paleontologist though you are, you show yourself unable to distinguish rocks from pebbles, for a pebble is what you reduce religion to." Now for the ruination.

Of Rocks and Pebbles

Gould says he cannot see what all the fuss is about, for (he tells us) "the conflict between science and religion exists only in people's minds, not in the logic or proper utility of these entirely different, and equally vital subjects." When tangle and confusion are cleared away, he says, "a blessedly simple and entirely conventional resolution emerges," which turns out (not surprisingly) to be his own. "Science tries to document the factual character of the natural world, and to develop theories that coordinate and explain these facts. Religion, on the other hand, operates in the equally important, but utterly different, realm of human purposes, meanings, and values."

Note that it is human (not divine) purposes, meanings, and values that Gould's "religion" deals with, but the deeper issue is who (in Gould's dichotomy) is to deal with the factual character of the nonnatural, supernatural world. No one—for to his skeptical eyes the natural world is all there is, so facts pertain there only. He has a perfect right to that opinion, of course, but to base his definitions of science and religion on it prejudices their relationship from square one. For it cannot be said too often that the issue between science and religion is not between facts and values. That issue enters, but derivatively. The fundamental issue is about facts, period—the entire panoply of facts as gestalted by worldviews. Specifically here, it is about the standing of values in the objective world, the world that is there whether human beings exist or not. Are values as deeply ingrained in that world as are its natural laws, or are they added to it as epiphenomenal gloss when life enters the picture?

That this *is* the real issue is lost on Stephen Jay Gould, but not on all biologists. Two years ago I was asked to speak to the evolution issue at the University of California, Davis, in a lecture that its office of religious affairs arranged. Several days after returning home I received a letter from the biology professor who teaches the evolution course on that campus. He said that he had come to my lecture expecting to hear things he would need to refute at his next class session but had been pleased to find little of that nature in what I had said. Enclosed with his letter was an article he had written in which he raised the question of what the evolutionary fuss was about. His answer was: "It is not about whether or not evolution is good science, whether evolution or creation is a better scientific explanation of the diversity of life, or whether natural selection is a circular argument. The fuss actually isn't even really about *biology*. It is basically about worldviews." *Rocks of Ages* could have been a helpful book if Gould had recognized this point, but now, having had my fun with Gould, I must admit that I have not been entirely fair to him. For he is quite right in saying that the position he advocates is "entirely conventional." That does not make it right, but it does exonerate Gould from having invented the mistake, which I quoted Appleyard as indicating a few pages back. "Separating our values from our knowledge of the world [is the chief way scientism] burns away ancient authorities and traditions."

From Warfare to Dialogue

Religious triumphalism died a century or two ago, and its scientistic counterpart seems now to be following suit. Here and there diehards turn up—Richard Dawkins, who likens belief in God to belief in fairies, and Daniel Dennett, with his claim that John Locke's belief that mind must precede matter was born of the kind of conceptual paralysis that is now as obsolete as the quill pen—but these echoes of Julian Huxley's pronouncement around midcentury that "it will soon be as impossible for an intelligent or educated man or woman to believe in god as it is now to believe that the earth is flat" are now pretty much recognized as polemical bluster. It seems clear that both science and religion are here to stay. E. O. Wilson would be as pleased as anyone to see religion fail the Darwinian test, but he admits that we seem to have a religious gene in us and he sees no way of getting rid of it. "Skeptics continue to nourish the belief that science and learning will banish religion," he writes, "but this notion has never seemed so futile as today."

With both of these forces as permanent fixtures in history, the obvious question is how they are to get along. Alfred North Whitehead was of the opinion that, more than on any other single factor, the future of humanity depends on the way these two most powerful forces in history settle into relationship with each other, and their interface is being addressed today with a zeal that has not been seen since modern science arose.

This could be in part because money has entered the picture (the Templeton Prize for Progress in Religion is larger than the Nobel Prizes), but it probably signals a change in our climate of opinion as well. Scientists probably sense that they can no longer assume that the public will accept their pronouncements on broad issues unquestioningly, and this requires that they present reasons. In any case, God-and-science talk seems to be everywhere. Ten centers devoted to the study of science and religion are thriving in the United States, and together they mount an expanding array of conferences, lectures, and workshops. Several hundred science-and-religion courses are taught each year in colleges and universities around the country, where a decade or two ago you would have had to dig in hard scrabble to find one; and every year or so new journals with titles such as *Science and Spirit, Theology and Science,* and *Origins and Design* join the long-standing *Zygon* to augment the avalanche of books—many of them best-sellers—that keep the dialogue between science and religion surging forward.

On the whole, this mounting interest is a healthy sign, but it hides the danger that science (I reify for simplicity's sake) will use dialogue as a Trojan horse by which to enter religion's central citadel, which is theology. That metaphor fails, however, because it carries connotations of intentional design. A hole in a dyke serves better. If a hole appears in a Netherlands dyke, no finger in the dyke is going to withstand the weight of the ocean that pushes to enter.

Colonizing Theology

To once have belonged to the enemy camp provides one with insights into its workings, and so (with apologies for the military language) I will claim that advantage here.

When I came to America from the mission field of China, my theological landing pad at Central Methodist College in Missouri was naturalistic theism, the view that God must be a part of nature, for nature is all there is. With modest help from John Dewey, Henry Nelson Wieman was the founder of that school of theology, and my college mentor was one of his two foremost protégés. Thus it was that when I arrived at the Divinity School of the University of Chicago to study with Professor Wieman, I was already as ardent a disciple as he had ever had. That lasted through my graduate studies, after which my resonance to the mystics converted me to their worldview.

At the time I am referring to (the middle of the twentieth century), Wieman's liberal naturalistic theism was giving its conservative rival—neo-orthodoxy, as founded by the Swiss theologian Karl Barth and captained in America by Reinhold Niebuhr—a run for the Protestant mind. Niebuhr won that round, but with Whitehead and his theological heir, Charles Hartshorne, naturalism has returned as Process Theology. Its philosophy of organism (as Whitehead referred to his metaphysics) is richer than Wieman's naturalism, and Whitehead's and Hartshorne's religious sensibilities were more finely honed, but Process Theology remains naturalistic. Its God is not an exception to principles that order this world, but their chief exemplar. God is not outside time as its Creator, but within it. And God is not omnipotent, but like everything in this world is limited. "God the semicompetent" is the way Annie Dillard speaks of this God.

Do we not see the hand of science—which process theologians point to proudly—in this half-century theological drift? In relating it to the concerns of this chapter, two questions arise. First, if we could have our way, would we prefer God to be fully competent or partially competent? Second, has science discovered any *facts* that make the first (traditional) alternative less reasonable than the second? If it has, science has vectored the drift and we must follow its lead. If no such facts have turned up, scientistic styles of thought are guilty of colonizing theology.

With this quick reference to the last fifty years, I turn now to the present.

The Tilt of the Negotiating Table

Because scientists at this point are negotiating from strength and would be happy to have things remain as they are, it is theologians who must take the initiative to get conversations going. I have already mentioned the ten or so religiously based institutes that are working at this job, and in these pages I shall confine myself to the two most prestigious of these, the Zygon Center at the University of Chicago, and the Center for Theology and the Natural Sciences at the Graduate Theological Union in Berkeley. In an informal division of labor, the Institute in Chicago publishes *Zygon,* the academic journal in the field, and the Berkeley Center mounts the conferences.

Who gets published in *Zygon* and invited to CTNS conferences? There is no stated policy, but an inductive scan suggests a bias against those who, first, criticize Darwinism; second, argue that the universe is intelligently designed; and third, accept the possibility that God may at times intervene in history in ways other than through the laws by which nature works. God may be believed to have created the universe and to operate within it, but God must not be taken to suspend at times its laws or to leave gaps in them that are divinely filled from outside. (That would give us a "God of the gaps," a deity who would be squeezed out when, as it is assumed will happen, science eventually fills those gaps.) In a word, miracles and supernaturalism generally are out. Those who honor the three mentioned proscriptions are welcomed in CTNS/*Zygon* doings; others are not.

Such at least is my reading of the matter. If the reading is basically accurate, the operative policy is pretty peculiar once one thinks about it. Three planks of the traditional religious platform have been removed by the pace-setting Berkeley/Chicago axis. (The religious platform I posit here is drawn from Hinduism and the Abrahamic religions. Buddhism and East Asia present complications that would be distractions in this discussion.) Why? The obvious answer seems to be that these planks do not fit the scientific worldview. I cannot speak for the governing boards of the two institutions and do not know if their policy here is tactical—to keep scientists from walking away from the negotiating table—or if it reflects a belief that science has discovered things that require that the traditional planks be dropped. I know the Berkeley team well enough to know that its members are sincere Christians who do not see themselves as capitulating to the scientific worldview if it is read in ways that exclude God. But the God they argue for is (1) the world's first and final cause, who (2) works in history by controlling the way particles jump in the indeterminacy that physicists allow them. This retains God, but in ways that supplement the scientific worldview without ruffling it.

The problem with this approach is that it overlooks the ghost of Laplace, who waits in the wings to announce that he has no need of the God-hypothesis. More serious is the procedural way things are going. The institutions that dominate the science-religion conversation do not consider the way they relate theology to science to be one possibility among others that merit hearings. They consider it to be the truth and believe that it needs to be understood if religion is to survive in an age of science.

Darwinism provides the clearest example of this monopolistic approach. That the issue of how we human beings got here has strong religious overtones goes without saying, and its founder and I are only two among millions who find the Darwinian theory (when taken to be fully explanatory of human origins) pulling against the theistic hypothesis. Among scientists themselves, debates over Darwin rage furiously, fueled by comments such as Fred Hoyle's now-famous assertion that the chance of natural selection's producing even an enzyme is on the order of a tornado's roaring through a junkyard and coming up with a Boeing 747. But when religion enters the picture, scientists close ranks in supporting Darwinism, with CTNS and *Zygon* right in there with them. To my knowledge, no one critical of the theory has been published in *Zygon* or been included in a major CTNS function.

Michael Ruse of the University of Guelph—a self-confessed bulldog for Darwinism—puts this colonization of theology by biology in perspective when he charges his fellow Darwinists with behaving as if Darwinism were a religion. Rustum Roy, a materials scientist at Pennsylvania State University, goes further. Half seriously, he has threatened to sue the National Science Foundation for violating the separation of church and state in funding branches of science that have turned themselves into religions. If these spokespeople are right and Darwinism has grown doctrinal, we have the curious spectacle of its colonizing not only theology but biology as well. I will close this chapter with an instance.

The 1999 conference on "The Origin of Animal Body Plans and the Fossil Record" was held in China because that is where a disproportionate number of fossils relating to the Cambrian explosion of phyla have been found. On the whole, its Western delegates argued that the explosion can be explained through a Darwinian approach, whereas the Chinese delegates were more skeptical of that. Jonathan Wells, of the Center for Renewal of Science and Culture at the Discovery Institute in Seattle, closed his report of the conference with an account that carries overtones ominous enough to warrant its being quoted in full:

> I'll end this report with one poignant anecdote about a conversation I had with a Chinese developmental biologist from Shanghai who recently returned from doing research in Germany. She told me that in China the general practice in education is to settle on an official theory and then teach it to the exclusion of all others. So far, she said, this has not happened in biology; since she herself is a critic of the idea that genetic programs control development, she dreads the possibility of being forced to teach the Darwinian line. But she fears that this may happen soon, and she and her colleagues believe their only hope is the willingness of western scientists to discuss competing theories and not descend into dogmatism. It depressed her to see at this conference how dogmatic American biologists had already become, and she pleaded with me to defend the spirit of free inquiry. The way she put it, the world is counting on you to do this.

CHAPTER 5

THE TUNNEL'S LEFT WALL: HIGHER EDUCATION

Turning now to the left wall of the tunnel, which is higher education, let us begin with the book that establishes this chapter's course. Again in this case (as in Chapter Three) there was no shortage of books to choose from. Page Smith's *Killing the Spirit: Higher Education in America* was a leading contender, but I have settled on George M. Marsden's *The Soul of the American University* because its subtitle, *From Protestant Establishment to Established Nonbelief,* comes close to telling the story in itself.

THE FLAGSHIP BOOK

The first American colleges were created to train clergymen, and it followed as a matter of course that a religious atmosphere pervaded their campuses. This atmosphere persisted for decades as the objectives of education expanded beyond the training of ministers. Only a century ago, almost all state as well as private universities and colleges held compulsory chapel services, and some required Sunday church attendance as well. Today, however, the once-pervasive presence of religion on campuses has all but disappeared. *The Soul of the American University* is not a lament for a lost golden age when the WASP establishment ruled. It does argue, however, that the addition of feminist and multicultural perspectives need not and should not have excluded traditional religious viewpoints, which can enrich the college curriculum without threatening sound scholarship and free inquiry.

The history of this matter is so familiar that I need only dub in some of the highlights of Marsden's thoughtful thesis, along with my own commentaries.

What Happened

American colleges were founded in a time of fervent national and moral idealism, and it would have been surprising if their founders had not viewed their practical concerns through religious lenses. Those lenses were evangelically Protestant, and the clergymen-presidents of the early colleges typically taught courses on the Bible and Christian doctrine and encouraged campus revivals. From the start, however, colleges recognized truths that could be reached by "natural reason" and without the help of revelation. Philosophy was the province of those truths, and natural philosophy (the early name for science) was the branch of it that dealt with nature. Early in Harvard's history one of its presidents quoted as the truly golden saying of Aristotle, "Find a friend in Plato, a friend in Socrates, but above all find a friend in Truth," and he went on to extol natural philosophy explicitly: "For what is Natural Philosophy, unless a system in which natural things are explained; and in which that hypothesis is certainly the best by which the greater part of natural phenomena are most fully and clearly explained. These things are to be sought and acquired."

So science and religion were allies at the beginning. But as the two subsequent centuries have unfolded, religion has steadily been pushed to the periphery. Colleges and their successor universities (I shall use the terms *college* and *university* interchangeably, for there is little difference between them save that of size) did not act in isolation here, however. By and large they simply kept pace with the advancing secularization of American society while doubling back in the twentieth century to confirm that secularization.

The most important cause of that increased secularization has been the progressive "technologizing" of the Western world in the name of progress, and universities have been key agents in that project. Scientists have been needed to discover new laws of nature, and engineers to put those laws to use. Everybody got into this act, not just universities and scientists, for from healthy bodies to microwave ovens and television sets, material goods are the most obvious trophies that life sets before us. It follows that nothing could have prevented the veritable explosion of science and engineering on campuses. Land-grant colleges were established explicitly to bolster the practical side of learning, but as science and engineering expanded by their own momentum in the older, more prestigious universities, the line between "state colleges" and universities all but disappeared. The most recent newcomers to the educational growth industry have been schools of business and management; for once we learn how to create products, the focus turns to mass production, advertising, and distribution, all of which lie within the province of corporations. Foreign students used to come to the West for science degrees, but a master's degree in business administration from the Harvard Business School is now said to be the most coveted prize.

Given the modern world, this burgeoning of science, technology, and schools of business on campuses was inevitable and in itself not inappropriate. It has exacted a price, however. The humanities and social sciences, which study *people,* have been elbowed to the sidelines.

I shall come back to this point, but several other social developments have affected the "feel" of the educational experience so much that they need to be mentioned before I resume the main concern of this chapter, which is the way the university shapes students worldviews.

- Bulging enrollments have turned universities into megaversities. In my college days we students were in and out of our professors' homes all the time. The logical limit of today's depersonalized education is courses that proceed entirely by Internet. One of my graduate students devised such a course. Unable to land a face-to-face teaching position, he seized the initiative and created a course on world religions, which the University of California Extension continues to offer for credit. In the five years it has been online, he has yet to lay eyes on a single one of his students.
- If burgeoning enrollments have depersonalized education, burgeoning knowledge has fragmented it. Renaissance men who knew something about everything that was to be known disappeared several centuries ago. Students now face a plethora of compartmentalized fields of knowledge. Uninstructed as to how they connect, students are given no sense of the whole, if indeed their instructors think a seamless fabric of knowledge exists.
- With rising tuition costs most students must now work while they learn, which leaves them tired much of the time.
- Vocational objectives have taken over. Higher education has always been a vehicle for social mobility, but now a college degree is needed simply to stand still and stave off the specter of the minimum wage.
- Two other developments in higher education are relevant to our discussion here, but these I place in a different category because their long-range effects promise to be enriching: the

ethnic complexion of campuses is changing, and women are more visible.

Doubtless there have been compensating gains in the four disturbing changes I noted, and it is always a mistake to underestimate the capacity of the human spirit to adapt to new circumstances. But my point in noting these social developments is really to put them aside. They needed to be mentioned because they affect the human spirit, but having taken note of them I return to the metaphysical tunnel this book is concerned with. Its metaphysics is naturalism and the point of this chapter is to note how the university's inattention (at best) to a reality that exceeds nature, and (at worst) its denial that such a reality exists, shape students' minds.

To restate the definition that has already been entered, naturalism is not materialism. Materialism holds that only matter exists. Naturalism grants that subjective experiences—thoughts and feelings—are different from matter and cannot be reduced to it, while insisting that they are totally dependent on it. No brains, no minds; no organisms, no sentience.

This is a prominent aspect of the scientific worldview, and what has happened to higher education is that it has been overtaken (or taken in) by it. There is a story to the effect that when, early in the twentieth century, a student went to Benjamin Jowett (then headmaster of Balliol College at Oxford University), distressed that he had lost his faith in God, Jowett thundered, "You will find it by nine o'clock tomorrow morning or leave this college." An apocryphal witticism, no doubt, but it highlights how times have changed.

Though I say that the scientific worldview has taken over, I must stress again that this has not been by design. The takeover is simply the culmination of the unconsidered outworkings in the university of the scientism that imbues modernism throughout. The incredible success of science acts like a magnet on the other departments of the university and causes them to ape its methodology. At the last social science colloquium I attended, the speaker (an economist) opened by asking if the social sciences are becoming more scientific. His answer was, "Not fast enough."

The Pull of Science on Other Disciplines

Because magnetic attraction is strongest at close range, it is not surprising that among other divisions of the university it is the social sciences that feel the pull of the natural sciences most strongly. As a social scientist himself, Robert Bellah has had to live with that pull throughout his career, and since he is exceptional in the clarity with which he recognizes the pull in question and the courage with which he protests it, I cannot do better than to turn this next section of the chapter over to him.

The Social Sciences

"The assumptions underlying mainstream social science," Bellah writes,

> can be briefly listed: positivism, reductionism, relativism and determinism. I am not saying that working social scientists could give a good philosophical defense of these assumptions, or even that they are fully conscious of holding them. I mean to refer only to, in the descriptive sense, their prejudices, their pre-judgments about the nature of reality. By positivism I mean no more than the assumption that the methods of natural science are the only approach to valid knowledge, and the corollary that social science differs from natural science only in maturity and that the two will become ever more alike. By reductionism I mean the ten-

> dency to explain the complex in terms of the simple and to find behind complex cultural forms biological, psychological or sociological drives, needs and interests. By relativism I mean the assumption that matters of morality and religion, being explicable by particular constellations of psychological and sociological conditions, cannot be judged true or false, valid or invalid, but simply vary with persons, cultures and societies. By determinism I do not mean any sophisticated philosophical view, but only the tendency to think that human actions are explained in terms of "variables" that will account for them.

Most social scientists, Bellah goes on to add, do not think of these assumptions as conflicting with the assumptions of religion. The assumptions are so self-evidently true that they are beyond contradiction. Religion, being unscientific, could have no reality claim in any case, though as a private belief or practice it may by some be admitted to be psychologically helpful for certain people. "Yet these assumptions conflict, and conflict sharply with every one of the great traditional religions and philosophies of mankind."

> Social science embodies the very ethos of modernity, Bellah continues, and for it there is no cosmos, that is, no whole relative to which human action makes sense. There is, of course no God, or any other ultimate reality, but there is no nature either, in the traditional sense of a creation or expression of transcendent reality. Similarly, no social relationship can have any sacramental quality. No social form can reflect or be infused with a divine or cosmological significance. Rather, every social relationship can be explained in terms of its social or psychological utility. Finally, though the social scientist says a lot about the "self," he has nothing to say about the soul. The very notion of soul entails a divine or cosmological context that is missing in modern thought. To put the contrast in another way, the traditional

> religious view found the world intrinsically meaningful. The drama of personal and social existence was lived out in the context of continual cosmic and spiritual meaning. The modern view finds the world intrinsically meaningless, endowed with meaning only by individual actors, and the societies they construct, for their own ends.

Because Bellah says exactly what I would have wanted to say—and says it with the authority of a respected insider—I will let him complete this section:

> Most social scientists would politely refuse to discuss the contrasts just mentioned. They would profess no ill will toward religion; they are simply unaware of the degree to which what they teach and write undermines all traditional thought and belief. Unlike an earlier generation of iconoclasts, they feel no mission to undermine "superstition." They would consider the questions raised above to be, simply, "outside my field," and would refer one to philosophers, humanists, or students of religion to discuss them. So fragmented is our intellectual life, even in the best universities, that such questions are apt never to be raised. That does not mean that they are not implicitly answered.

Psychology

Psychology has fractured. Experimental psychology comes close to being an exact science, but most of our minds and selves are beyond its pale, and this leaves clinical, or depth, psychology to pick up the residue. Whereas experimental psychology deals with people as objects, clinical psychology approaches them as subjects. The differences in methodology that these approaches require are so great that the two camps have difficulty communicating with each other and must be considered separately here.

In experimental psychology, Pavlov's salivating dogs, J. B. Watson's behaviorism, and B. F. Skinner's updated version of the two are obviously within the gravitational pull of the hard sciences. To these can be added stimulus-response theory generally: in holding that actions followed by rewards are repeated, Thorndike's pacesetting Law of Effect is thoroughly mechanical. Every decade or so it gets a facelift, but its explanatory limitations are built into it, so the spotlight has moved to cognitive psychology. Because that will be considered in a later chapter, I will skip it here and turn directly to clinical, or depth, psychology.

Perhaps the most telling fact here is the university's stonewalling of models of the self that make more room for the human spirit than the orthodox Freudian view does. The chief of those alternative models are the ones proposed by first, C. G. Jung; second, humanistic and transpersonal psychology; and third, Asian religions, all of which have proven so useful to practicing therapists that they have spawned Jungian Institutes, the existentially oriented Association for Humanistic Psychology, and the Association for Transpersonal Psychology. All three entities are flourishing and have led to the founding of accredited programs for training therapists outside the university. (The California Institute for Professional Psychology, Saybrook Institute, and Pacifica Institute in Santa Barbara are three in my own backyard.) But their proven usefulness has not gained them entrance to the university.

It takes no great feat of mental gymnastics to recognize the hand of Freudian orthodoxy in this closure. Daniel Goleman, former behavioral science editor for the *New York Times,* says that Freud's depiction of the human self is the closest the West comes to having a model, and he does not think well of it. It is more pessimistic than the alternative models that the extramural psychologies work with. (Psychiatrists Roger Walsh and Dean Shapiro have pointed out that the index to the collected works of Sigmund Freud contain over four hundred entries for pathology and none for health.) It is also more deterministic—existential psychology emerged to challenge this aspect of Freudianism. But because Freudianism is uncompromisingly materialistic and purports (in the face of Adolph Grunbaum's and Frederick Crews's demonstrations to the contrary) to be scientific, it fits better with the prejudices of today's university.

The Humanities

Advocates of the human spirit, the humanities were traditionally the heart of higher education. Today they are neither its heart nor its center. Having been replaced at the center by professionalism and science, the humanities are now outlying provinces—in enrollments, budgets, and prestige, all three. A 1998 article in *Harvard Magazine* noted that the number of undergraduate degrees given in the humanities has plummeted since 1970, both absolutely and proportionally. On average, humanists receive the lowest faculty salaries by thousands (or tens of thousands) of dollars, their teaching loads are the heaviest, and their time allotted for research is the least.

This can be seen as a takeover by technology and vocationalism, but it can just as well be seen as abdication; humanists have renounced their post as moral mentors. Emerson argued that "the whole secret of the teacher's force lies in the conviction that men are convertible, and they are. They want awakening, [and for that purpose they need teachers] to get the soul out of bed, out of her deep habitual sleep." That was long ago. Today humanities instructors no longer wrestle with the purpose of human existence and the correct ordering of the soul. "The sad news," according to Robert Scholes, "is that teachers of literature [his field] are in trouble because we have allowed ourselves to be persuaded that we cannot make truth claims but must go on 'professing' just the same." Carl Woodring adds dryly, "Literature is useful for a skeptical conduct of life."

This skepticism, which all of the humanities now foster, is effected by two main thrusts, deconstruction (which comes close to being the heart of postmodernism) and the hermeneutics of suspicion. Both call for a brief word. First, deconstruction.

In the game of name-that-song (or that age, or century), *postmodern* is the best that historians have been able to come up with as a label for the second half of the twentieth century and what follows. Pointing (as the word does) to no more than a stretch of time, it has no positive content of its own, but deconstruction stepped in quite early to mend that lack. As was noted in the opening chapter of this book, postmodernism began as a movement to question the grand narratives of the Enlightenment and human progress and has gone on from there to question all worldviews. Charging that systems (both social and conceptual) are coercive, it takes as its task their dismantling.

I have never heard deconstruction presented as an extension of Gödel's theorem into philosophy and literary criticism, but *undecidability* is the bottom line for both. From Aristotle to Turing, mathematicians had tried to establish systems that are complete. Gödel smashed that dream. His famous Incompleteness Theorem states that in a formal system satisfying certain precise conditions, there will always be at least one undecidable proposition—that is, a proposition such that neither it nor its negation is provable within the system. Jacques Derrida's denial of any single meaning in a text sounds like a direct extension of this. The interminable activity of interpretation that follows from that denial—an ongoing sparring match between alternative plausible explanations in the hope of generating new ideas, new values, new understandings of the world and ways we might respond to it—is the stock-in-trade of deconstruction. Whether the new ideas are better than the original ones is seldom squarely addressed—and indeed is almost beside the point where breaking-open-to-allow-for-alternatives is itself the name of the game.

As I wrote this page, David Carson's *2nd Sight: Grafik design after the end of print* crossed my desk. Its message as displayed on its cover read: "creativity is unusual stuff. it frightens. it deranges. it's subversive. it mistrusts what it hears. it dares to doubt. it acts. even if it errs. it infiltrates preconceived notions. it rattles established certitudes. it incessantly invents new ways. new vocations. it provokes and changes point of view." The point for this chapter is the way (in modes such as that one) deconstruction contributes to the erosion of belief that Marsden sees universities now abetting, for beliefs that are built out of cards that are endlessly reshuffled are not intended to remain stable.

As for the hermeneutics of suspicion, I begin with Steven Weinberg's report of an elderly friend of his who (at the prospect of his impending death) says he draws some consolation from the fact that when that event arrives he will never again have to rush to his dictionary to look up what the word *hermeneutics* means. It means *interpretation,* but *hermeneutics* sounds more imposing.

The hermeneutics of suspicion is an interpretive device that attacks theses not head-on but indirectly, by innuendo. Is it proposed that X is the case? Suspicious hermeneutics responds, not by arguing that it is or is not the case—indeed, it does not engage the claim at that level at all—but by changing the subject to the unacknowledged motives that (it alleges) are the real reasons for advancing the claim. In rhetoric courses this is known as the fallacy of psychologizing, and the hermeneutics of suspicion battens on it. If the attack is mounted from a Marxian angle, the real reasons for making the claim (it is charged) are the claimant's class interests. From the Freudian angle, repressed aggression or libido are cited as the real cause. In everyday examples, the claimant is accused of wanting to make a name for himself, or to be a provocateur.

For the notion of truth, the hermeneutics of suspicion has been a disaster. In Foucault and much of postmodernism generally, truth comes close to being no more than a power play. Wilfred Cantwell Smith reported that although *veritas* remains enshrined on the insignia of Harvard University, the word does not once appear in a statement on the aims of undergraduate education that its faculty took two years to hammer out shortly before Smith retired.

The hermeneutics of suspicion has its place, for motives regularly figure in human doings. I will go so far as to admit that this entire first half of my book can be read as an extended investigation into the way motives we were not conscious of have worked to cause us to pin our hopes excessively on science. But I do not make such muckraking my supreme concern. My supreme concern is the nature of things, to which the second half of this book is devoted.

The slackening of that concern is what produces the nonbelief Marsden is troubled by. Robert Bellah endorses Marsden's thesis emphatically. The deepest indictment of today's university, Bellah says, is that it erodes not just religious belief but all beliefs other than those of science. I ran into an unusually vivid instance of the phenomenon in my own backyard recently. At a Labor Day block party a newcomer recognized me and (giving his interest in philosophy as his reason) asked if we could lunch together. When that happened, he catalogued a string of way-stations that he had moved through—psychedelics, India, Rajneesh, Da Free John (a familiar list)—and then came to the point. "My problem," he said, "is that I am convictionally impaired. I can believe something for a year or two, and then it dissolves and I start searching again." *Convictionally impaired*—Marsden's "nonbelief" in a nutshell, and also a testament to Philip Rieff's charge that the essence of modernity is "the settled hunting down of all settled convictions."

Philosophy

Outside the Western world, philosophy and theology can hardly be separated, and in the West too they were partners through and beyond the Middle Ages. Clement described Christianity as the confluence of two rivers, Athens and Jerusalem, and Thomas Aquinas forged the medieval synthesis by adding Aristotle's metaphysics to the foundations of Christian theology. In the Middle Ages, philosophy was the handmaid of theology, and (with Hume as the lone dissenter) God remained the kingpin in the great modern metaphysical systems down through Hegel. Hegel's was the last important theistic philosophy, however, for though German idealism and nineteenth-century romanticism slowed the advance of the scientific worldview temporarily, early in the twentieth century logical positivism swept the two aside. Linguistic philosophy slowed positivism in the third quarter of that century, but the century closed with its materialistic premise back in place. I remind the reader of John Searle's earlier-quoted assertion that professionals in philosophy now accept some version of materialism because they believe that it is the only philosophy consistent with contemporary science.

That God has no place in such philosophy goes without saying, but what counts more is the fact that God's absence is now so taken for granted that it is hardly noticed. It used to be that while theists and atheists differed in their conclusions, both sides considered the question important, but that common ground has collapsed. The confrontational iconoclasm of Bertrand Russell and

Jean-Paul Sartre has given way to the atheism of apathy, indifference, and unconcern.

With respect to the human spirit, philosophy's compliance with the gravitational pull of science is only half of the story. The other half is its reinforcement of that pull by actively pushing itself away from religion. As long as metaphysics and moral philosophy were high on its agenda, it allowed itself to be housed in departments of philosophy and religion, but when logic displaced those priorities, cohabitation became uncomfortable. As Bertrand Russell's dictum that logic is the essence of philosophy took hold, the ability to follow completeness proofs for formal systems via symbolic logic replaced foreign languages as a graduate requirement, and increasingly philosophy has found itself with less and less in common with religion.

Self-esteem entered the picture also, for religion's low status in the university caused philosophers to resent being associated with it and to demand their own departments. Richard Rorty suggests that present-day philosophy may be playing out the gloomy vision of Henry Adams, who (a century or so ago) regarded the new religion of science as being as self-deceptive as the old-time religion had been, and believed that its "scientific method [was] simply a mask behind which lurk[ed] the cruelty and despair of a nihilistic age." The recent founding of the Society of Christian Philosophers may seem like a counterexample to what I am saying, but the advice that aspiring members hear from their mentors is, Don't write your dissertation on philosophy of religion. Write it on something else. When you get a job, *then* you can do philosophy of religion.

Religious Studies

When state universities and colleges were created, it was initially assumed that the constitutional separation of church and state prohibited the teaching of religion in public institutions. Around the middle of this century, however, a distinction was drawn between teaching objective facts about religion and proselytizing for it, and this paved the way for religious studies departments to spring up on most campuses.

This has not served the human spirit as much as might have been expected, for when higher education adopted the European model of the university, it took over its way of studying religion, which was as positivistic as its way of studying other subjects. (More on this important topic soon.) Auguste Comte had laid down the line: religion belonged to the childhood of the human race. It is good to know facts about childhood, but retention of its outlook shows that you are childish yourself. This did not get religious studies off to a promising start. The discipline's founding fathers, who continue to be revered as its giants—linguist Max Müller, anthropologist Emile Durkheim, and sociologists Max Weber and Karl Mannheim—were either agnostics or atheists. Müller confessed to being religiously "unmusical," and Mannheim pretty much spoke for the crowd when he said, "There is no Beyond. The existing world is not a symbol for the eternal. Immediate reality points to nothing beyond itself."

These early prejudices remain in place. Once the world was divided up along scientific and nonscientific lines, sociology became (as Peter Berger has pointed out) a more formidable enemy of religion than science is, for it claimed jurisdiction over "social man," defined as people in the totality of their experience. History had paved the way for the takeover by insisting on religion's historical character: religions arise not from divine incursions into the world but from historical circumstances, and they are therefore relative. Freud spun out the psychological variant of this theme by arguing that religion is a projection of human needs and desires, a view all the more sinister because of the unedifying character of the needs and desires Freud postulated.

Sacred myths and texts are the heart of religion, and its adherents accept them as revealed. Having fallen from heaven, so to speak, they bring news of a reality that exceeds and surpasses our everyday world. Religious studies (whose methodologies are of a piece with those of the humanities and social sciences generally) cannot accept this claim at face value. I will let two biblical scholars tell the story—one as it bears on the New Testament, the other as it applies to the Torah. For the New Testament, Marcus Borg writes:

> To a large extent, the defining characteristic of biblical scholarship in the modern period is the attempt to understand Scripture without reference to another world. Born in the Enlightenment, which radically transformed all academic disciplines, modern biblical scholarship has sought to understand its subject matter in accord with the root image of reality that dominates the modern mind. "Rational" explanations—that is, "rational" within the framework of a one-dimensional understanding of reality—are offered for texts which speak of "supernatural" phenomena.
>
> The major sub-disciplines which have emerged in biblical scholarship are those which can be done without reference to other levels of reality: studies of the way the biblical writers redacted the tradition which they received, the form and functions of various literary and oral genres, the rhetorical development of early Christian tradition expressed in the texts, etc. All share in common the fact that they focus on the "this worldly" aspects of the texts.

For the Hebrew Bible, Arthur Green has this to say:

> The emergence of *Wissenschaft* [the science of history, in the broad, European meaning of the word *science*] brought forth the

bifurcation between the study of Torah as a religious obligation and the forging of scholarly research into a surrogate religion of its own. We are forced to "bracket" for the purposes of teaching and research our faith in God. The methods by which religion is studied in the university are those of history and philology (in the humanities) and anthropology, psychology, and sociology (in the social sciences). Their impact has been to discount Torah as a divine creation. A scholar who submitted an article to the *Journal of the American Academy of Religion* or the *Journal of Biblical Literature* assuming that Scripture was quite literally the Word of God would be a laughing stock.

From Nonbelief to Disbelief

The Soul of the Modern University: From Protestant Establishment to Established Nonbelief has been invaluable in charting the course for this chapter, but the two preceding subsections in my own text show it stopping short of the mark. The modern university is not agnostic toward religion; it is actively hostile to it. It countenances spirituality as long as it is left undefined; I have never encountered students who did not think that they had a spiritual side to their nature. But organized, institutionalized spirituality (which is what religion comes down to) is not well regarded on campus. In a follow-up book to the one I have been using, Marsden gets to this point by writing, "Younger scholars quickly learn that influential professors hold negative attitudes toward open religious expression and that to be accepted they should keep quiet about their faith."

History helps to place this prejudice in perspective. We know that universities evolved in Europe from cloisters; the word *college* initially referred to cloisters of monks who needed to know how to read in order to perform their offices. And as has been noted, in the New World the first colleges were primarily seminaries for training ministers. But when colleges in the course of becoming universities loosened or severed their church connections, they needed a new identity (a new model, if you will), and the German universities, then the most prestigious in the world, were ready at hand. They were positivistic to the core, and (because they have retained their place as the model for the American university) it is important to understand the militant secularism that is built into the word *positivism.*

Positivism is the philosophical position usually associated with the name of the nineteenth-century French sociologist Auguste Comte, its most influential popularizer. The term *positive,* as he used it, carries the sense of something given or laid down, something that (because of its givenness) must be accepted at face value and without need of explanation. As its negative corollary, the word warns against the attempts of theology and metaphysics to go beyond the evidential world in the (vain) hope of discovering first causes and ultimate ends. All genuine knowledge is contained within the boundaries of science. Philosophy remains useful in explaining the scope and methods of science, but nothing good can be said for theology. Religion belongs to the childhood of the human race, and philosophy too must abandon the claim to have means for attaining knowledge that science lacks. Indicative of this blatant hostility toward religion is the fact that when (at the close of the eighteenth century) the German theologian Friedrich Schleiermacher gave a series of lectures titled *On Religion* at the University of Marburg, he subtitled his lectures *Speeches to Its Cultured Despisers.*

As a philosophical position positivism has collapsed, but the anticlericalism it built into the eighteenth-century German university remains in place in the American university today. Force of habit explains this in part, but rivalry also enters the picture. Having won its autonomy from the church, the university has become the church's rival for the mind of our times, and rivals seldom have the fairest pictures of their opponents' positions.

This is not a pleasant topic. I do not recall, in the innumerable science-religion discussions I have been party to over the years, ever having heard it laid on the table. But it is a fact of life, and to face it squarely I reach for John Kenneth Galbraith's description of the rivalry that the New Deal (devised by Franklin Delano Roosevelt to pull the United States out of its Great Depression) created between Washington and the business community. "Ideology had a role," Galbraith concedes; "the free enterprise system needed to be defended, and the business community saw itself as its defender. This was the approved motivation. But deeper and more powerful was the sense of lost position, of prestige passing to Washington from New York, Pittsburgh and Detroit. It was that as much as ideology which mobilized the businessmen against Franklin D. Roosevelt."

"The sense of lost position" and "of prestige passing" to others are the telling clauses here. In his book *Name-Dropping,* Galbraith generalizes their point in a paragraph that is so germane to the university's hostility toward religion that I shall quote it in full:

> Two motivating forces exist in any economic system: one is the desire for money; the other is the need for prestige. The pursuit of money—income—is widely accepted. But for the business community, prestige is also deeply important and something not to be shared. The only acceptable economic policy [from the business standpoint] is one that accords front rank to the corporate executive or the financier. An active government, like Roosevelt's, all too obviously challenges the basis for business esteem and self-esteem. Better to suffer some loss of income [as

business did in the Great Depression] than to see this prestige—the right of leadership—impaired or invaded.

All that needs to be changed in those words to explain the prejudice against religion on today's campus (and the disbelief in matters religious that it fosters in students) is to replace the rivalry between business and government in Galbraith's account with the rivalry between the science-dominated university, on the one hand, and religion, on the other, for the university is deeply invested in its claim to control knowledge. The parallel is not exact, for business saw itself threatened by the New Deal, whereas religion is in no position to threaten today's science-dominated university. But it *has* threatened education in the past, and memories die slowly. Moreover, off campus (in society at large) the competition between the two sides for the public mind continues apace.

The Ineffectiveness of the Theological Response

The university's assault on religion placed theologians in a difficult position. They needed to counter it to make a place for their concerns; at the same time, however, they did not want to withdraw from the intellectual life of culture, and higher education had firmly established itself as the primary institutional center for developing the knowledge on which a modern scientific-technological society depends. More important, theologians were themselves products of the university and to an appreciable extent had taken on its coloration.

If I were to choose a flagship for the present short section, it would be Douglas Sloan's *Faith and Knowledge: Mainline Protestantism and American Higher Education,* for it tells in detail the story that I will compress into several paragraphs. The strategy that twentieth-century theologians adopted to deflect the university's assault on religion was to argue for two kinds of truth. (We anticipated that strategy in the preceding chapter through Stephen Jay Gould's concurrence with it.) On the one hand are the truths of knowledge as these are derived from science and from discursive, empirically grounded reason. On the other hand are the truths that faith, religious experience, morality, meaning, and value put forward. The latter are not grounded in knowledge. They arise out of a blend of feeling, intuition, ethical action, communal convention, folk tradition, and mystical experience.

The strength of the twofold approach is that it helps to keep alive important dimensions of human experience and meaning that the dominant view of knowledge cannot encompass. It has, however, a fatal weakness. While resisting in certain ways the modern mindset, at a deeper level it buys into the basic cleavage in the modern theory of knowledge that Jacques Monod points to in the passage I quoted at the end of Chapter Two—the split between subject and object, fact and value, theory and practice, science and the humanities, and (for religion) faith and knowledge. And, of course, the balance between the two realms is unequal. The domain of faith, meaning, and values is constantly placed on the defensive and undercut by the incursions of a narrow, positivistic knowledge, along with its accompanying materialistic worldview. Not being grounded in the reality that is generally recognized to be potentially *knowable,* the object of faith, ethics, and art stands in constant danger of becoming epiphenomenal and only derivatively real. The crisis of faith in the modern world derives from the cognitive disparity between these two views of truth. Twentieth-century theology illustrated that disparity; it did not correct it.

One more occurrence deserves notice before this chapter closes.

The New Professionalism

The emergence of the American university in the second half of the nineteenth century brought with it a revolution in our understanding of intellectual life. Where those universities most revealed their spirit was in the manner in which they accommodated science and secularism, freed themselves from the religious orientation that guided the old colleges, embraced curiosity as a value in itself, and enshrined reason as the driving force in intellectual life.

This new spirit showed itself especially in the new professionalism, which reorganized old professions (theology, medicine, and law) and spawned new ones (business administration, journalism, veterinary medicine, forestry, and the like). The old professionalism took liberal studies seriously because they made human beings their central concern. The new professionalism studies *things,* and it raises questions not about humanity's ultimate role and the responsibilities that go with that role but about whether X or Y is the better way to go about achieving some immediate, restricted end. This is a difference not of degree but of kind. In enshrining instrumental knowledge and making it central, the burgeoning universities turned vocations into professions. In the process, not only was the focus on life's purpose and meaning lost, but the focus on the human *qua* human was attenuated, for it reduced human beings to instruments for the advancement of this-worldly knowledge.

Conclusion

I round off this chapter on education with two quips that are biting in their content but expressed in ways that keeps them from sounding bitter.

The reviewer of a recent spoof on midwestern colleges of agriculture—a book titled *Moo U*—opens by saying that the book is of course a satire, and then proceed to excuse that fact by asking how else is it possible to write about today's university.

The second observation was delivered by the art historian A. K. Coomaraswamy. Imported from India to establish the Asian wing of the Boston Museum of Fine Arts, he is said to have remarked that several decades as an immigrant had convinced him that it takes four years to acquire a college education in America and forty to get over it.

CHAPTER 6

THE TUNNEL'S ROOF: THE MEDIA

Regardless of their profession, most intellectuals are profoundly socialized by their formal education. This means that universities put the finishing touches on the minds of those who go forth to rule America. Small wonder, then, that the secularism and anticlericalism of those universities have spread to blanket our cultural life. Polls consistently show that the majority of Americans say they believe in God, but it would be a mistake to see that statistic as reflecting the standing of religion in public life. Peter Berger's justly famous quip has caught on: "If India is the most religious country on our planet, and Sweden is the least religious, America is a land of Indians ruled by Swedes." The next chapter will document his point as it relates to law; this one shows how the media reflect it.

THE FLAGSHIP BOOK

Edward J. Larson's *Summer for the Gods: The Scopes Trial and America's Continuing Debate over Science and Religion* is my flagship here. Written by a historian of science and professor of law at the University of Georgia, and published by Harvard University Press in 1998, it concerns itself with a single phenomenon: the media's handling of the 1925 Scopes trial in Dayton, Tennessee. Larson's analysis of that event is so revealing that I will devote the first half of this chapter to reporting it and defer generalizations about the media to its second half.

Inherit the Wind

Until historians started looking into the matter and their findings began to trickle out to the public toward the twentieth century's close, if you asked virtually any American what happened in the Scopes trial in Dayton, Tennessee, during the sweltering summer of 1925, the living-room answer would have come less from the facts of that case than from myths about it that were spun by a highly successful motion picture, *Inherit the Wind,* which was based on the equally successful Broadway play with the same name. Starring Frederic March and Spencer Tracy, who played William Jennings Bryan and Clarence Darrow, respectively, the film flipped reality like a pancake and elevated itself to mythic stature.

The bare facts from which the film takes off are generally known. (Because the film was based on the play, I shall refer to them interchangeably.) In 1925, Tennessee enacted a law against teaching evolution, and the American Civil Liberties Union (ACLU) advertised for a biology teacher to test its constitutionality. The ACLU had hoped to keep the trial to the free-speech versus majoritarianism issue—majoritarianism argues that the majority has the right to determine outcomes—but when William Jennings Bryan and Clarence Darrow entered the fray, it was inevitable that the science-religion issue would be central. It is that issue that the play and the film pick up on, with science cast as the knight in shining armor battling ignorant, bigoted, backward-looking religionists. The audience is told that the film is not "history" but a "historical film," but this does not keep the screenplay from coming across as if it were history. It tells in its own way the story of the "monkey trial," in which John Scopes was convicted of violating the law against teaching evolution. Under Stanley Kramer's theatrically brilliant and blatantly partisan reenactment of the trial, the film all but replaced the trial itself in the public's memory. Writing at the time, Irving Stone said that the film had dealt the deathblow to fundamentalism, which the film did not distinguish from religion generally. If time has proved Stone wrong, no credit goes to *Inherit the Wind.*

The Play's the Thing

Inherit the Wind opens with Leslie Uggams singing "Give Me That Old-Time Religion" in dirgelike, drumbeat cadence as three stern-faced officials and a preacher in a small Southern town march into a schoolroom to arrest a biology teacher who is rumored to be teaching evolution. We next see this teacher, John Scopes (there is no point in cluttering my account with the stage names that the film gives the protagonists), being visited behind bars by his lovely fiancée, who is caught between her love for her courageous, principled, soon-to-be husband and her love for her Bible-thumping father, who regards Scopes as the devil himself. This is a standard ploy for getting an audience on the side a film wants it on, but as *history* it is pure fiction. Romance played no part in the trial, and Scopes was never incarcerated. Bryan (the chief prosecutor) argued that he should not be sentenced at all, and when the judge said that such leniency was not within his jurisdiction, Bryan immediately offered to pay the minimum fine of one hundred dollars that the judge meted out.

The opening gambit sets the course for the film as a whole. A stage instruction notes that it is essential that the town be always visible in the background, and it is invariably shown as hostile to Scopes and his advocate, Clarence Darrow. When Bryan arrives at the Dayton train station, he is given a hero's welcome and an evening banquet, whereas the platform is empty when Darrow arrives and a hostile citizenry lines the sidewalks as he makes his solitary way to his hotel, its stony silence broken only by a high school girl who screams, "Fiend!" Nothing resembles history here. Contrary to the oppressive presence that the film turns Dayton into, it was a friendly and tolerant town; it was in a holiday mood for the trial and enjoying the worldwide attention it was getting. No mob demonstrated loudly outside the nonexistent jail chanting, "We'll hang John Scopes to the sour apple tree," and not even the flamboyant atheism of the trial's star reporter— H.L. Menken of the *Baltimore Sun*—induced him to say something as irresponsible as, "Hooligans of the world unite; you have no one to burn but your intellectuals."

The actual facts are these. In the hope of reversing three decades of declining population, Dayton's town fathers saw in the ACLU's search for a biology teacher to test the legality of the Tennessee law a golden opportunity to put Dayton back on the map. That its high school biology teacher was ill and incapacitated posed no problem; the football coach and general science teacher, John Scopes (who had been drafted to complete the biology course) would do. In the actual trial, Scopes testified that as substitute teacher in the second half of the course he had learned more biology from his students than they had learned from him, for at least they had had six weeks of instruction from someone who knew something about the subject. None of this is in the film.

The town fathers' strategy exceeded their wildest hopes. Over two hundred reporters alone poured into Dayton, and the trial turned out to be the first in America to receive international coverage.

Liberties of the sort I have mentioned lace the film throughout, all of them designed to make the point that science is reasonable, forward-looking, and tolerant, whereas religion (equated with fundamentalism) is bigoted, closed-minded, and backward-looking. Despite Bryan's explicit assertion on the stand that he read the six days of creation allegorically, he is depicted as a dyed-in-the-wool biblical literalist. What's more, the humanitarian side of his case is totally ignored. Bryan was first and foremost a passionate humanitarian. He was an irrepressible evangelist for social reform, and social Darwinism (which would soon be discredited) was then in its heyday. Bryan had seen the survival-of-the-fittest theory used to defend the robber barons in America, and in Germany to justify the brutal militarism that led to World War I. This had led him to believe that "the Darwinian theory represents man as reaching his present perfection by the operation of the law of hate, the merciless law by which the strong crowd out and kill off the weak."

One would never guess from the film that it was this perception of Darwinism that mostly drew Bryan's fire, for the film focuses on Darrow's making a fool of Bryan via his (inaccurately portrayed) beliefs on evolution and scoring thereby a thumping victory for science. The reporters who covered the trial saw things differently. They took the confrontation to be the opening skirmish in a battle between religious fundamentalists and religious modernists that would continue. The film shows Bryan collapsing and dying on the stand under the weight of Darrow's withering cross-examination, which—this much is accurate—had nothing to do with the case but everything to do with Darrow's aim to ridicule Bryan and biblical literalism. ("I made up my mind to show the country what an ignoramus he was," Darrow wrote to Mencken after the trial, "and I succeeded.")

In actuality, Bryan did not consider himself beaten at all. He spent the following days issuing statements to the press and preparing a fifteen-thousand-word stump speech that would continue his battle against Darwinism and Darrow. He did (as it turned out) die five days after the trial, but not because his spirit was broken. Referring to Bryan's legendary love for food, Darrow said he died of a busted belly.

I am writing as if *Inherit the Wind* acted alone in misrepresenting the Scopes trial, but the distortions started building a year or two after the trial itself. If a single book were to be credited with launching them, it would be Lewis Allen's *Only Yesterday,* which was published in 1931 and sold over a million copies. A racy book about the Roaring Twenties, it perpetuated the idea that Bryan was a fundamentalist who believed in the literal Genesis, which we have seen was not the case. Allen turned the trial into a contest between Bryan's and Darrow's opposing views of religion and evolution, completely omitting both Bryan's humanitarianism and his majoritarianism. "The real issue," Bryan had said on the stand, "is not *what* can be taught in public schools, but *who* shall control the educational system."

In all, Allen's treatment of the trial left the impression that it amounted to a triumph of reason over revelation, and this became the received version of the matter. By the time of the McCarthyite Fifties, even historians of the stature of Richard Hofstadter were citing the trial as an expression of the dark, anti-intellectual forces in America. Almost half a century was to elapse before this stereotype of the trial was clearly recognized to be such and efforts were initiated to correct it. In a piece titled "A Visit to Dayton," published in *Hens' Teeth and Horses' Toes,* Stephen Jay Gould called attention to misrepresentations of the Scopes trial's larger significance, and several other historical studies (Ray Ginger's *Six Days or Forever: Tennessee v. John Thomas Scopes,* and Edward Larson's initial look into the subject, *Trial and Error: The American Legal Controversy over Creation and Evolution,* which prepared the ground for his definitive study—the flagship for this chapter) back Gould up on that point.

Poetic License

Art has rights of its own, of course. It has to select and highlight to keep its storyline clear, and along the way it may have to pit good guys against bad guys. Be it said, therefore, that the revealing fact about *Inherit the Wind* is not that it takes liberties. Even its gargantuan licenses can be excused (if one wishes to excuse them) by calling them poetic. The revealing fact about the production is the message it delivers, and the way to place that message in perspective is to try to imagine the roles in *Inherit the Wind* reversed. In today's climate of opinion, can we imagine Hollywood using the Scopes trial as the basis for a story that cast William Jennings Bryan in the role of the hero and Clarence Darrow the villain?

Edward Larson does not charge that *Inherit the Wind* was spun from thin air; intellectual intolerance was certainly an important issue in the case. What Larson does (in addition to flagging the factual errors in the film) is to restore attention to aspects of the trial that are commonly overlooked and that continue to reverberate today in discussions about the place of science and religion in public schools. The fear Bryan voiced during the trial—"that we shall lose the consciousness of God's presence in our daily life if we must accept the theory [Darwinism] that through all the ages no spiritual force has touched the life of man and shaped the destiny of nations"—reads as if it could have been uttered yesterday. And in substance it *was* uttered virtually yesterday, in the recent brouhaha over Darwinism in Kansas.

KANSAS UPDATE

I would not have devoted the space I have to *Inherit the Wind* if I did not see it as the most graphic index I know of the way the media handle religion in our time. Media coverage of the Kansas Board of Education's August 1999 decisions regarding evolution confirms that impression, so I will tack that case on as an update. My local newspaper, *The San Francisco Chronicle,* followed the pattern across the country in titling its editorial on the decision "A Vote for Ignorance," but if we look below the surface we find that it was the media's coverage of the decision that raised the ignorance level of our nation.

If that sounds like an irresponsible charge, I ask the reader whether the following facts about the case come as a surprise. To the extent that they do, they reveal the press's shortcomings in reporting the happening.

Contrary to the impression the media gave, the Kansas decision actually *increased* its public schools' emphasis on evolution. The old science standards (in effect since 1995) devoted about 70 words to biological evolution, whereas the new one increased that to about 390 words. While that is short of the approximately 640 words the Kansas Science Education Standards Writing Committee wanted, it is still a fivefold increase over what had been on the books before.

Word counts don't tell the whole story, of course, but the 390 words approved by the school board included many of the provisions the committee recommended. The board adopted verbatim the committee's summary of Darwin's theory, which read:

> Natural selection includes the following concepts: 1) Heritable variation exists in every species; 2) some heritable traits are more advantageous to reproduction and/or survival than are others; 3) there is a finite supply of resources available for life; not all prog-

eny survive; 4) individuals with advantageous traits generally survive; 5) the advantageous traits increase in the population through time.

The board mandated that Kansas students be tested on this summary of Darwin's theory of natural selection—a summary that it would be hard to improve on. It also required students to understand that "microevolution . . . favors beneficial genetic variations and contributes to biological diversity," listing finch-beak changes as an example of this concept.

So what is so bad about Kansas, and why the uproar? The problem lies in the school board's refusal to adopt two of the proposals that its science committee would have liked to have had included.

First, it declined to require students to understand that microevolution leads to macroevolution—the origin of new structures and new groups of organisms. And second, it did not require students to elevate biological evolution into a "unifying concept" of science, on a par with such concepts as "evidence" and "form and function." But it is hard to regard those declinations as votes for ignorance when professional biologists themselves do not agree on those points.

The press's bias in reporting the Kansas affair comes to light not only in the way it covered the event, but also in the way it did not. I refer to the symposium on the decision that Washburn University in Topeka mounted in the wake of the uproar. As this was (to my knowledge) the only responsible academic discussion of the case that took place—responsible in that both sides of the controversy were allotted equal time—one would have thought that journalists would have seen it as an opportunity to add depth to the story, but not so. As far as I have been able to discover, outside Topeka the press ignored the event.

In doing so it withheld from the nation a telling fact—that the biology department at the Washburn University abstained from the discussion. What does its abstention do to the image of science as grounded in free and open discourse?

The General Picture

There is a scene near the end of *Inherit the Wind* that seems in present context as if it had been scripted to lead from the film itself to the broader issue of the media's treatment of religion generally. A character identified as Radio Man enters the courtroom carrying a large microphone. He explains that the microphone is connected by direct wire to Station WGN in Chicago. He then proceeds to report to the nation at large what is happening in the courtroom. William Jennings Bryan, famed as an orator with a huge voice, attempts to speak into the microphone. He fumbles with the new gadget, but his voice is loud enough to be picked up anyway and his words are conveyed to the public. In Bryan's final tirade, however, the program director in Chicago decides that his speech has aired long enough and Radio Man breaks in to announce that the station is returning to its Chicago studio for a musical interlude. The stage directions describe this as Bryan's final indignity.

Phillip Johnson of the University of California Boalt School of Law extracts from this scene an issue that he calls "Who owns the microphone?" The microphone (that is, the news media in general) can cancel anything Bryan might say by simply turning off his mike. Being himself a critic of Darwinism's exaggerated claims, Johnson ties this point to his own experience. In today's media atmosphere, he says, he finds it practically impossible to get newspapers to acknowledge that there are scientific problems with Darwinism that are quite independent of what anybody thinks about the Bible. A reporter may seem to get the point during an interview, but after the story goes through the editors, it almost always comes back with the same formula: creationists are trying to substitute Genesis for the science textbook.

My own counterpart to his report comes from my field of world religions. Several years back the religion reporter for one of the nation's leading newspapers flew to the Bay Area to interview me for a profile she intended to write. After the usual questions concerning my background, formative influences, human interest items, and my opinions on a variety of topics, she got around to religious conflicts. I told her that they tend to be more political than religious, and that triggered a mini-lecture that ran roughly like this:

In ethnic conflicts involving religion, religions provide the warring parties with their respective identities, but it does not follow that the differences in those identities are the cause of the conflict under review. In the way logicians put this point, multiple identities are a *necessary* condition for conflict, but not its *sufficient* condition. The case is the same as it is with people. For there to be a fight there must be distinct parties, but the plurality of the parties does not require that they fight. It provides the conditions for friendship as well as for hate.

That states the matter abstractly, I explained to the reporter, and then I offered an example to concretize the point. Several years ago, when the eyes of the world were on Bosnia, I happened to catch an evening news clip in which a reporter was questioning a woman in a Serbian village. The conversation went like this:

REPORTER: Are there any Muslims in your village?
WOMAN: No.
REPORTER: What would you do if there was one?
WOMAN: We would tell him to leave.
REPORTER: And if he refused?

WOMAN: We would kill him.
REPORTER: Why?
WOMAN: Because that's what they did to us four hundred years ago.

That pointed exchange pretty much says it all as far as religious conflict goes, I told her. Differences in religious beliefs—the sphere of religion proper—are not the primary causes of the problems; Serbs could not care less about what Muslims *believe*. It is the memory of atrocities unavenged that generally fuels the fires.

Not always, to be sure. When a religion *enters* history, its distinctive, defining beliefs are in conflict with those of its progenitors and neighbors and are thus perceived as threats. Was Jesus the Messiah or was he not? Was the Hindu caste system acceptable or, as the Buddha argued, not? Was Muhammad a prophet in the line of Abraham, Moses, and Jesus or was he an imposter? Questions such as these really *are* religious questions, and they led to wrenching, bloody battles as the budding faiths wrested their independence from their parents in something like world-historical adolescent rebellions. But once religions wring their independence from their parents and establish their own identities, I explained, it is political issues more than doctrinal differences that create problems. Furthermore, political leaders often use religion for their political ends.

The reporter heard me out and then said, "I think I understand very clearly what you've said, but my editor doesn't. What he wants from me is terrorism and holy wars, preferably jihads. If it bleeds, it leads." And sure enough (I speak for myself now), that is predominantly what the media give us when it comes to religion.

Having mentioned Bosnia, I will mention a second example in the same vein that involved me personally. It occurred in the 1970s when the Middle East was in flames. On the day that American hostages were taken in Lebanon, Americans could think of nothing else, and in the afternoon—I was teaching at Syracuse University at the time—I received a call from a colleague in political science. His course on the Middle East would be meeting that evening, and would I come to explain the role of religion in the Middle East crisis?

Collegiality led me to agree, but I can still remember my heavy tread as I made my way across campus to his classroom, for I knew that what I would say would disappoint him and his students. Experience had taught me that he hoped I would be able to point to doctrinal differences that could help to explain the conflicts in the region—most deeply those between the Muslims and the Jews—whereas what I would have to tell the class was that compared with the burning issue of land, theological niceties were pedantic. Truth to tell, the religious differences between those two faiths are so small that Muhammad could hardly have been more surprised when he discovered that the Jews and Christians of his day did not accept him as an addition to their own prophetic lines.

I return to the general picture. It must frustrate camera reporters no end that the human spirit is invisible. Being unphotographable, it never appears on the evening news. This leaves reporters having to cope with what can be visually portrayed, which (pertaining to religion) is its overt spin-offs. And here, we (the general public) enter the act with our addiction to violence. If a pro-life advocate shoots an abortionist doctor, it is certain to hit the front page of every newspaper in the country. Meanwhile, millions of ordinary citizens will on that day have given some thought to their souls through prayer, meditation, Bible-reading, and the like—activities that reach into the depths of the soul where the switches are thrown between kindness and cruelty, hope and despair. This passes without mention.

That is where the matter begins, but not where it ends. Reporters are taught as the art of their craft to keep their personal opinions out of their stories—"Nothing but the facts, ma'am; nothing but the facts." But the dictum does not apply when their opinions reflect the skepticism of our reigning ethos. E. J. Dionne of the *Washington Post* illustrates this with an anecdote concerning a dilemma that he once faced while on assignment in Africa. He was covering the pope, and while waiting in the rain for the papal plane to arrive, he fell into conversation with a Catholic who was standing beside him. Dionne voiced his concern about the weather, but his neighbor told him not to worry, that it would stop raining when the pope arrived. "How do you know that?" Dionne asked. "The rain doctor said so because the pope is blessed" was the reply.

Dionne would not be telling that story if things had not transpired exactly as the rain doctor had predicted: the sun came out, and the pope conducted the mass in a brilliant, rain-washed world. Whenever he tells this story to journalism students, Dionne asks how they would have handled it. He finds the predictable consensus to be *as a coincidence.* But what, he then asks, does that do to the principle of nothing but the facts?

I return to Peter Berger's characterization of America as a land of Indians ruled by Swedes. Christopher Lasch's book *The Revolt of the Elites* is devoted to this issue, and a paragraph from it is worth quoting here. That Americans generally *say* they believe in God means nothing, Lasch says, for

> public life is thoroughly secularized. The separation of church and state, nowadays interpreted as prohibiting any public recognition of religion at all, is more deeply entrenched in America than anywhere else. Religion has been relegated to the sidelines of public debate. Among elites it is held in low esteem—something useful for weddings and funerals but otherwise dispensable. A skeptical, iconoclastic state of mind is one of the distinguishing characteristics of the knowledge classes. Their commitment to

> the culture of criticism is understood to rule out religious commitments. The elites' attitude to religion ranges from indifference to active hostility. It rests on a caricature of religious fundamentalism as a reactionary movement bent on reversing all the progressive measures achieved over the last several decades.

Intellectuals typically present religion as comforting people with the agreeable illusion that they are the center of the universe, the object of God's loving-kindness and rapt attention, Lasch continues. But (in a paragraph that is not easy reading but is important enough to quote) Lasch points out that it is just this illusion that the most radical form of religious faith relentlessly attacks. Thus,

> Jonathan Edwards distinguishes between a "grateful good will" (the root of religious feeling, as he understood it) and the kind of gratitude that depends on being loved and appreciated—the kind of gratitude, in other words, that people might feel toward a creator presumed to have their interests at heart. "True virtue," Edwards wrote, "consists, not in love of any particular beings nor in gratitude because they love us, but in a union of heart to being in general." Man has no claim to God's favor, and a "grateful good will" has to be conceived, accordingly, not as an acknowledgment of the answer to our prayers, so to speak, but as the acknowledgment of God's life-giving power to order things as he pleases, without "giving any account of his doings."

This view of God, Lasch concludes, bears no resemblance to Freud's benign father figure conjured up by childlike human beings out of their unconscious need for dependence. Freud (whom intellectuals tend to look to as the authority on this matter) assumes that religion answers the need for dependence, whereas Edwards extols those who self-confidently disclaim any such need. Indeed, such persons find it galling to be reminded of their personal dependence on a power beyond their own control.

The line that runs from Christopher Lasch's observations to the media is not circuitous, and the object of this chapter is to point out how straight it is. Before I close, however, I want to insert a few paragraphs about advertising. They will be brief, because advertising is a social institution, and worldviews, not society, are this book's beat. But just as in the preceding chapter I found it necessary to take passing note of certain structural changes in the university (because their effects on students' spirits have been so pronounced that not to mention them at all would have looked like an oversight), so likewise here.

Who Pays the Piper?

It is advertising that in large part governs the media, because it is advertising that pays the media's bills. Insofar as it takes it upon itself to inform people of products they are unaware of that could enhance their lives, advertising performs a valuable service, but it would be naive to assume that advertising agencies see that as their mission. Persuasion rather than information is their object. Industrial societies may actually require that advertisers be persuaders, for technology has become so efficient at mass-production that the problem has shifted from production to consumption—moving products out of warehouses to forestall glut. This makes advertising and it's near-equivalent marketing the decisive links in capitalism's feedback loop.

That paragraph positions advertising in its social context, but how does advertising impact the human spirit? I have not yet defined *spirit,* but when I get around to doing so, character will be one of its important components, and advertising leaves its marks on character.

Imagine three different scenarios that follow from finding a wallet lying on the sidewalk. In the first, the finder pockets the money, throws the wallet into the nearest Dumpster, and happifies himself with the thought that this is his lucky day. In the second, the finder agonizes over the choice but keeps the money. She does, however, post the wallet (with its credit cards) back to the owner. The third finder does not hesitate. She seeks out the nearest phone, informs the owner that his wallet has been found, and refuses the reward he offers her.

Everyone would agree that the character of the wallet's finder rises as we progress through this sequence. The question is how to *acquire* the character of the third finder.

The first step is to establish ourselves as moral agents. This involves learning to captain our desires instead of being their slaves. If we fail to achieve that upper hand, a Japanese adage reports the consequence: "First the man takes a drink, then a drink takes a drink, and then drink takes the man." As to which of our desires should be reinforced, it is those that will benefit us in the long run and will further the public good.

Advertising works against all of these moral needs. It presses for immediate gratification and things that will benefit oneself instead of the general public.

Conclusion

It has been this chapter's object to document the obvious—that the minds of journalists have been forged in the academy and shaped by its secular hammerings. Garry Wills calls journalists on that point. In their eyes, he says, their secular outlook justifies them in ignoring the 120 million people or so in America who regularly practice their religion. "It is careless," Wills continues, to keep misplacing such a large body of people. Nonetheless, every time religiosity catches the attention of the intellectuals, it is as if a shooting star had appeared in the sky. One could hardly guess that nothing has been more stable in our history, nothing less budgeable than religious belief and practice. Religion does not shift or waver. The attention of its observers does. Public notice, like a restless spotlight, returns at intervals to believers' goings on, finds them still going on, and with expressions of astonishment or dread, declares that religion is undergoing some boom or revival.

Peter Jennings (senior anchor and editor of ABC's *World News Tonight* as I write these lines) used that statement of Garry Wills to close an address he delivered at the Harvard Divinity School several years ago on "The Media's Challenge in Covering Religion." To it he added these words of his own: "We must stop treating religion as if it were like building model airplanes, just another hobby, not really a fit activity for intelligent adults. The sooner we do, the sooner we will have a greater grasp of our nation."

My own conclusion for this chapter derives from something Saul Bellow said during the three weeks he spent at Syracuse University in the early 1980s. At the press conference that the university mounted on his arrival, one of the reporters asked him, "Mr. Bellow, you are a writer and we are writers. What's the difference between us?" Bellow answered, "As journalists, you are concerned with news of the day. As a novelist, I am concerned with news of eternity."

I anticipate the second half of this book when I add to Mr. Bellow's remark the assertion that the third millennium will be well served if the existing gap between these two writing professions is narrowed.

CHAPTER 7

THE TUNNEL'S RIGHT WALL: THE LAW

Trying to relate the human spirit to laws is tricky business. For one thing, laws keep changing; every important decision adds a precedent that future decisions must build on. Philosophers used to say that to characterize Bertrand Russell's philosophy you had to look at your watch—when did he write the piece in question. Something like that pertains here.

A second problem is that opinions differ as to what the constitutional doctrine of the separation of church and state primarily intends. Is the intent to protect churches from governmental interference or to protect politics from religious pressure groups? Underlying both of these problems is the fact that there is no way to keep church and state separate. They have always jostled each other, and always will. Let me say, therefore, that the right wall of the tunnel that this chapter describes is the wall as fashioned in the second half of the twentieth century. Whether developments that have occurred since this book went to press suggest that we are emerging from the tunnel or moving deeper into it, readers must decide for themselves.

THE FLAGSHIP BOOK

For the final flagship book that I shall be using, I have chosen Stephen Carter's *The Culture of Disbelief: How American Law and Politics Trivialize Religious Devotion.* Something odd happened an hour ago that relates to the subtitle of that book. A typical day for me begins (after a few minutes of hatha yoga) with the reading of a passage from one of the world's enduring scriptures. This morning (just before sitting down to start this chapter on the law) I found myself reading, in the Gospel According to Luke, "Woe to you, lawyers. For you load people with burdens hard to bear and you yourselves do not lift a finger to ease them. . . . You have taken away the key of knowledge. You did not enter yourselves, and you hindered those who were entering."

Dismiss this as coincidence, see it as Carl Jung's synchronicity, or think of it as God working in disguise—be my guest. Something Boris Yeltsin once said, however, encourages me to report the incident. Asked by an American reporter about his religious beliefs, he said that naturally as a Marxist he had none. As he settled back, apparently awaiting the next question, he did a double-take and reverted to what he had just said. "No, I am not religious," he reaffirmed, "but I *am* superstitious."

Back to business. If the denunciation of lawyers I quoted were left unqualified, I would want no part of it. But as I noted, this chapter considers their actions in the second half of the twentieth century, and in that period they have things to account for, as Stephen Carter's book indicates.

Carter, who teaches law at Yale University, explains that he wrote his book because he had noticed an increase in the marginalization of religion in public life during the thirty years of his career and wanted to look into the law's role in the decline. As he told an interviewer when his book was published,

> in an earlier era, although there was never as healthy a respect for religious pluralism as there should have been, I do think there was a healthy respect for what counted as religion. People might have been somewhat limited in their visions of what counted as religion, but there was a respect for it, and I think this was true right across the political spectrum and up and down the social and economic ladder. That has changed. There is less respect for religion, less of an appreciation of it as an important force that can genuinely be the motive force in people's lives without being somehow a symptom of something neurotic. That's what's been lost.

The legal system's contribution to that loss fits snugly into the story of the twentieth century that I have been telling. As the century unfolded, the dominant, liberal-rationalist culture increasingly imposed on the public "a common rhetoric that refused to accept the notion that rational, public-spirited people can take religion seriously." Not only did the courts buy into that rhetoric; increasingly they arrogated to themselves the power to reinforce it. Carter's criticism of this is not shrill. For the most part he simply urges that rulers treat religious concerns more respectfully than they have taken to doing. But he does come down hard on a 1990 United States Supreme Court decision, *Employment Division v. Smith*, which stripped the Native American Church of its constitutional rights. Because that decision claimed two years of my working life (two of its most rewarding years but more on that later), I will use it for the showpiece in this chapter, a role parallel to that played by *Inherit the Wind* in the chapter preceding.

Employment Division v. Smith

For whatever reasons (perhaps because the issue was too hot to handle), the framers of the U.S. Constitution left religious matters to the states. This was the clear intent of the First Amendment: "Congress shall make no law respecting the establishment of religion or the free exercise thereof." Two centuries later, *Employment Division v. Smith* flew squarely in the face of this stipulation and stood it on its head. The Oregon State Supreme Court had ruled that one of its citizens, Alfred Leo Smith, was entitled to belong to his Native American Church, and the U.S. Supreme Court overruled that decision. Since the story that produced that ruling is not widely known outside legal circles, I will summarize it.

Born on the Klamath reservation, Alfred Smith was (at the age of eight) taken from his parents and placed in a Catholic parochial school. His entire formal education took place in boarding schools. He talks of the consequences:

> Those were difficult times for me. I was separated from my family and stripped of my language, my culture, and my identity, and eventually I became an alcoholic. At the age of thirty-six I stopped drinking and began a life of recovery through Alcoholics Anonymous. Fifteen years later I was introduced to my first sweat lodge ceremony. That was the beginning of my introduction to the way my ancestors had lived, and to this day I receive spiritual guidance through the Native American Church.

After his recovery Smith developed Native American programs for alcohol and drug abuse. His final job in that field was in Roseburg, Oregon, where he was hired to help develop services for Native American clientele. Things proceeded smoothly until one Friday afternoon his superior called him into his office and asked him if he was a member of the Native American Church. When Smith said that he was, his superior asked if he took "that drug" (i.e., peyote). "No," Smith replied, "but I do partake of my church's sacrament." His boss told him that peyote was illegal and that he was unwilling to have a lawbreaker on his payroll. The following Monday his boss summoned him back into his office and asked if Smith had gone to his church over the weekend. When Smith said that he had, the boss again asked if he had taken "that drug." When Smith answered as before—"No, but I did take my church's sacrament"—he was terminated (along with another member of the church who worked in the same agency).

Native Americans are not well schooled in standing up for their rights. (I once heard Daniel Inouye, chair of the Senate committee on Indian Affairs, say in the course of a congressional hearing, "It does not please me to report that in the over eight hundred treaties that the United States has signed with the Indians, the United States has broken every one of them, while making sure that the Indians lived up to their side of the agreements.") Alfred Smith, however, proved to be a notable exception. He did not ask to be reinstated, but he did ask to receive the benefits he had earned, and (when they were denied him) he carried his case through the Oregon courts. These swung back and forth on the issue until, six years later, the supreme court of the state vindicated his claim. But then Oregon's attorney general referred the decision to the U.S. Supreme Court, which overturned the Oregon ruling.

That the highest court's decision violated both the letter and the spirit of the Constitution—its letter, because the First Amendment forbids the federal government to take actions that would interfere with the free exercise of religion; its spirit, because the intent of the amendment was to turn religious issues over to the states—has already been remarked, but the ethics of the case also warrants mention. I speak with some passion here, for (as I mentioned in introducing this case) I was drawn into its consequences. That the U.S. Supreme Court singled out for oppressive action the weakest, most oppressed and demoralized segment of our society is travesty enough—first we take away their land, and then we turn on their final refuge and take away their religion as well—but we need also to consider the nature of peyote, the sacrament of the Native American Church on which the case turned. Not, though, before I indicate how I became involved in the case.

One of my students, James Botsford went into Native American law, and the morning after the Smith decision he phoned me to ask if I wanted to get involved in the movement to win back the Native Americans' rights. I said that I did, and consequences ensued. The court's decision drew from the Native Americans a response the likes of which they had not theretofore marshalled. Under the inspired direction of one of the twentieth century's great Native American leaders, Reuben Snake ("your humble serpent," he said to me when we were introduced), they mounted the Native American Religious Freedom Project, which involved virtually all of the three hundred–odd Indian tribes in the United States. The judiciary having deserted them, they did an end run around it and carried their case to Congress.

Because congressional representatives are sensitive to the wishes of their constituents, the Native American coalition saw a need to educate the public about the issue in question. After they produced a documentary film, *The Peyote Road,* Reuben decided it needed a book to accompany it, and he assigned me the task of writing (or as things turned out as the book came together, editing) it. Things moved faster than we had anticipated, and by 1994 Congress had passed Public Law 103-344, the American Indian Religious Freedom

Amendments, which restored to the Native American Church its constitutional rights. That turned the book into a celebratory account of the Native Americans' victory over the highest court of the land. Co-edited by Reuben Snake and titled *One Nation Under God: The Triumph of the Native American Church,* it tells a story to inspire freedom-loving peoples everywhere.

With this autobiographical connection included, I go back now to the agent that provoked the *Smith* decision, peyote. Peyote is illegal in the United States at present. It is classified as a Schedule One drug—right up there with crack, heroine, and cocaine—and the mistake begins right there, for peyote is a harmless cactus to which addiction is virtually impossible and to which not a single misdemeanor (let alone crime) has ever been traced. When we place this record alongside the ravages of alcohol the picture becomes surrealistic. Because alcohol is the sacrament of the dominant religion of the land, it passes muster; but peyote is "their" sacrament, so it does not. One of the ironies of this drama-packed story is that (thanks to the Native Americans' vision and determination), the Native American Church, formerly singled out for disenfranchisement, is now by congressional act the only church in the United States that enjoys explicit legal protection.

Other churches have not fared as well.

The Religious Freedom Restoration Act

Employment Division v. Smith sent shockwaves through the churches of the land, for while the Native American Church was its direct target, its ramifications did not impact that church alone. Watchdogs for the major churches had been following the *Smith* case closely, seeing consequences in it for religious freedom in general: "If it's them today, tomorrow it could be us." So it was that, the day after the U.S. Supreme Court's decision, the largest coalition of religious bodies ever to unite in a common cause—some seventy-five in all—entered a brief asking the court to reconsider its decision, which it refused to do.

The churches had reason to be concerned, for no one had expected the provisions of *Smith* to be so far reaching. Through hundreds of federal and state cases relating to American religious freedom in the last two hundred years, the phrase "compelling state interest" had emerged as the test for state intervention. Unless the state could prove that there was a compelling need to intervene, it was not entitled to do so. *Smith* lowered that threshold to a "rational basis."

To support this retreat from the established threshold, Justice Antonin Scalia (who wrote the decision) argued that America's religious diversity had proliferated to the point where religious freedom was a "luxury" that a pluralistic society could no longer "afford." In withdrawing the "compelling interest" standard, the court also removed from First Amendment protection the entire body of criminal law. This, in effect, rewrote the First Amendment to read, "Congress shall make no laws except criminal laws that prohibit the free exercise of religion." (Put more simply, *Smith* mandated Congress to disregard the First Amendment if the law being considered is classed as a criminal law.) Finally, the court suggested that the First Amendment does not protect the free exercise of religion unless some other First Amendment right, such as speech or association, is involved. This, of course, makes religious freedom irrelevant, for those other rights are independently protected. Milner Ball, professor of constitutional law at the University of Georgia, said at the time that "after *Smith,* there is a real and troublesome question about whether the free exercise clause has any real practical meaning in the law at all. When you need the First Amendment, it won't be there. Or at least, that is the way the *Smith* case has left the law."

I have already referred to the consternation that the *Smith* decision awakened in the religious community, and it sprang into action immediately. With the strong support of President Clinton, the coalition of churches succeeded in getting Congress to pass the 1993 Religious Freedom Restoration Act, which restored the "compelling interest" phrase as the standard that government agencies needed to meet before they could interfere in religious affairs. The churches breathed easier, but only for three years, for in 1997 the Supreme Court struck down that act on grounds that Congress had overstepped its constitutional authority in passing it.

Marginalizing Religion

I return to Stephen Carter.

Legal landscape is neither monochrome nor unchanging, so (as would be expected) there have been deviations from the general course that I have traced. But on balance the courts have been "transforming the Establishment Clause in the First Amendment [which, to repeat, states that "congress shall make no law respecting an establishment of religion] from a guardian of religious liberty into a guarantor of public secularism," Carter writes. It is getting to the point where you can't pray out loud anymore, especially in the North, without drawing five kinds of scholars and an ironist or two, a legal scholar has quipped.

Carter acknowledges the liberals' point that the American ideal is threatened when religious power mixes too intimately with political power. He argues, however, that the greater threat comes when the church is no longer kept merely separate but is forced into a position of utter subservience, its voice disregarded in the great public discussions (or even disqualified from joining them). American liberalism is showing toward religion an increasing hostility, Carter argues, and the consequent culture of disbelief threatens more than religious misfits—Moonies, Hutterites, and the like. The real danger is that citizens in general will accept the culture's assumption that religious faith has no real bearing on civic responsibility. Should that happen, prevailing cultural mores will have a higher claim on us than do privately held convictions of conscience, however arrived at.

Our political discourse accommodates itself to this wall of separation. Civil religion ("The Battle Hymn of the Republic" at inaugurations, "In God We Trust" on our currency) reinforces rather than refutes Carter's point, for shallow deference to religious forms works to trivialize and "domesticate" authentic faith. Rather than elevating politics to religion's level—"Let justice flow down like waters, and righteousness as a mighty stream"—they preempt religion for political ends.

Carter waxes theologically metaphysical here, and it is good that he does, for that form of presentation has taken a backseat for a while. The heart of Carter's argument is that faith, in "a powerful sentience beyond human ken," necessarily carries with it a spirit of opposition to the prevailing culture, and this makes the tension between church and state as proper as it is inevitable. The Berrigan brothers were right, as were Thoreau, Martin Luther King, Jr., Mahatma Gandhi, the Amish, the Hutterites, and (in their more acculturated ways) the Mennonites and Quakers (or Friends) in their uncompromising resistance to war.

"At their best," Carter quotes David Tracy, "religions always bear extraordinary powers of resistance. When not domesticated as sacred canopies for the status quo nor wasted by their own self-contradictory grasps at power, religions live by resisting." And

(James Carroll adds) the state lives by being resisted. The genius of the American system is that the separation clauses of the Constitution weave this possibility into its very fabric. That is why this country has for more than two centuries thrived by changing.

Because religion's perspective is rooted not only outside its institutions and outside the national code, but also outside history and time itself, citizens whose religion really matters to them provide an inexhaustible source of the energy needed for human renewal. How? By enacting, in Carter's phrase, "the role of external moral critic, and an alternative source of values and meaning." Without the churches' determined support, the civil rights movement of the sixties could not have succeeded, and without their equally determined opposition we would have had American troops in Guatemala and El Salvador a decade later, Robert Bellah reminds us. *This,* at root, is why Carter deplores the culture of disbelief and law's contributions to it.

Handling Creationism

This chapter began with a case study, *Employment Division v. Smith,* and I will become concrete again as I draw it to a close. Nowhere do the traditional and scientific worldviews collide more sharply than on the question of how we human beings got here, which is why Darwinism erupts so frequently in this book. In the preceding chapter I noted how the media are handling it, and here the point is how the issue is being handled by the law.

Courts require that children attend school, and in public schools only science's answer to the question of human origins may be taught. From Tennessee (1925), through Arkansas (1982), to Louisiana (1987), attempts have been made to clear a space where (implicitly if not explicitly) God might be brought into the picture, but on alleged constitutional grounds the courts have consistently denied God that space.

I do not argue that (given the way the cases alluded to were framed) the courts ruled wrongly. My point, rather, is that we do not have the pieces in this picture positioned aright. How they *should* be positioned I am reserving for later, but that something is wrong in the present setup seems quite apparent.

Reduced to simplest terms, courts rightly assume that theism is a religious position, while wrongly assuming that atheism is not. It will be countered that atheism is not *taught* in public schools, which is true if teaching is taken to proceed explicitly only and not implicitly as well, but no educational theorist thinks that the two forms of instruction can be separated completely. If God is omitted from accounts of human origins, students will take that absence as implying that God has no place in the picture.

Specifically, when in 1987 the U.S. Supreme Court struck down Louisiana's statute requiring that creation-science be taught alongside evolution-science, Justice William Brennan argued for the majority (in *Edwards v. Aguillard*) that the state law was an unconstitutional "establishment of religion" because the legislature's purpose "was clearly to advance the religious viewpoint that a supernatural being created mankind." The phrase *creation-science* muddies the water by not distinguishing between six-day and long-term divine creation (we are talking billions of years here). But Justice Brennan's ruling is stated in broad terms that cover both readings: schools may not teach that a purposeful supernatural being (God) has played a part in our being here. This is a clear case of marginalizing. Religious claims are not squarely faced for their truth or falsity. Rather, they are eased out of the picture by classifications; in this case, theism is religious, while its alternative is not. This is supposed to reflect a national policy of neutrality, but the move is anything but neutral when the effect is to exclude important ideas and public policies from national scrutiny and debate.

If the courts were to say that the naturalistic worldview is true and (as the logical implicate of this) theism is false, it would be exposed as having scrapped the Establishment Clause. What it has done instead is to create legal categories that *contain* the theistic worldview, sealing it from public discussion and relegating it to the private sphere. This is as if in a debate the judge were to decide for the negative, not because its arguments were stronger but because the affirmative's arguments were ruled out of order. The consequences are far-reaching. "There is no doubt that in developed societies education has contributed to the decline of religious belief," Edward Norman writes in *Christianity and the World Order;* and Martin Lings targets the primary cause of this. "More cases of loss of religious faith are to be traced to the theory of evolution than to anything else."

Conclusion

Pro Deo et patria, for God and country, we say, and the conjunction is appropriate. Resistance does not presume radical separation of church and state any more than does identification. It presumes a dynamic interaction of the one realm with the other. But the last half of the twentieth century has shown how little we know what the right interface between the two comes down to in any specific confrontation. The good guys and the bad guys have switched sides unpredictably, and we can be sure that they will continue to do so. The state claims the prerogatives of the church at its peril, and the opposite is equally true; we need only recall the revival-like Republican convention of 1992, which sent out warning signals by nailing into its platform planks lifted from the religious right.

Neither side enters this fray from a position of superiority. If faith properly enables religious citizens to resist the unjust policies of government, it does so because it has first enabled those citizens to resist the dark sides of themselves.

PART 2

THE LIGHT AT THE TUNNEL'S END

Having devoted the first half of this book to describing the tunnel that modernism shunted us into by mistaking scientism for science, I turn in this second half to the future. Is light appearing at the tunnel's end? Are we stalled on a siding? Are we—this I do not think is the case—continuing to move deeper into the tunnel because we have not yet reached its center?

These are important questions, and the chapters ahead take them on if for no other reason than that our interest in the future is part of what makes human beings interesting. Those questions are not, however, the main object of this book's remainder. After they have been toyed with, the closing chapters of the book, beginning with Chapter Thirteen, settle into the most useful way to prepare for the future, which is to be clear about features of the religious landscape that do not change. History is unpredictable,—crystal balls are always a mystery—but a map that registers unchanging aspects of the terrain can orient us whatever comes our way. In the process it will help to show us why religion matters.

CHAPTER 8

Light

Science can prove nothing about God, because God lies outside its province. But as I devoted a chapter in my *Forgotten Truth* to demonstrating, its resources for deepening religious insights and *enriching* religious thinking are inexhaustible.

I begin with light. Light is a universal metaphor for God, and what science has discovered about physical light helps us to understand (more profoundly than even the spiritual giants of the past could do) why light is uniquely suited for that role. If Einstein could say at one point in his career that he wanted for the rest of his life to reflect on the nature of light, surely the question warrants a short chapter to set in motion the second half of this book. Light is different. It is strangely different. And paradoxically different. All three of these assertions hold for God, as does a fourth. Light creates.

The Physics of Light

Uncanny as they are, the basic features of Einstein's Special Theory of Relativity have worked their way into our common stock of knowledge. The speed of light—186,000 miles a second—is an unvarying constant, and everything else in the physical universe adjusts to it. Seated in their stalled cars at a railroad crossing, impatient drivers see the train whizzing past them, while its passengers see the cars flying past *their* windows in the opposite direction. That relativity concerns space; but physics locks space, time, and matter together like pieces of a jigsaw puzzle, so the relativity just mentioned turns up in time and matter as well. If you are hurtling through space, time (your watch) slows down. On a bicycle the slowing is unnoticeable, but if you flew a trillion miles in a fighter plane and landed at six in the evening according to your watch, the clocks in the airport you departed from would read seven. And the closer your plane came to traveling at 186,000 miles per second, the slower its clocks would run, until (if that speed were reached) the plane's clocks would stop. As for matter, the mass of an object that is in motion increases until, should it attain the speed of light, that mass would be infinite as measured by an unmoving observer.

Now let us spin all this around and look at it from light's point of view. Imagine yourself sitting on a particle. On that single "piece" (or quantum) of light you are going nowhere. You are weightless. There is neither time nor space, nor are there separated events. If from the earth it is one hundred light-years to a star, from your position on your quantum of light the star and earth are not separated at all. Moreover, it would seem as if the world were pouring out of you, you and your fellow photons, because light creates. It pumps power into the spatio-temporal world. This is most obvious in the process of photosynthesis, where the immaterial light flowing from the sun is transformed into the earth's green carpet of vegetation. Plants absorb light's immaterial energy-flow and store it in the form of chemically bonded energy. If we look beneath biochemistry at nature's foundations, we see that light's creativity "comes to light" there through its early appearance in the sequence that produces matter in its successive stages. (The phrase "comes to light" is not a pun. Everywhere in recorded history light doubles for intelligibility, comprehension, understanding, and—underlying all of these—conscious awareness. This metaphorical use of light reveals its protean power.) Situated on the cusp between the material and immaterial realms, photons (as was just noted) are not subject to our usual ways of understanding the physical universe.

Everything that was compressed into that preceding paragraph is strange, so it will not hurt to restate its content. Space? Remember that seated on light—a photon—you are going nowhere. Time? Time does not exact from photons the toll that it does elsewhere; how could it when clocks stop at the speed of light? As for matter, photons have neither the rest-mass nor the charge that material particles have. In lay language (which cannot get into *quarks* and *gluons* and other specks that sound like *Star Trek* aliens), these material particles derive from energy, which—using the word in its broadest sense—I am calling light. In addition to having rest-mass and charge, these derivative particles are also subject to time, so they are clearly material in all of those respects. Still, they are not *completely* material, for no definite position in space can be assigned to them. Atoms are more material than particles are for being locked into both space *and* time, but even they are not as "fallen" as molecules are, for isolated atoms are free to absorb and release energy to a much greater extent than atoms that have combined to form molecules, which are almost completely imprisoned in the determinism of our inanimate macro-world.

If (in some such way as I have described) light produces the physical universe, it is also responsible for its permutations. Quantum mechanics tells us that the essence of every interaction in the universe is the exchange of quanta of energy. A single quantum is the smallest packet of energy that can ever be exchanged; Planck's Constant is its measure. It is quanta of photons that change molecules in the act of photosynthesis and that excite atoms in the retina of our eyes to enable us to see. The exchange of light maintains our universe from the level of atoms and molecules on up.

What this all comes down to is that the two great epochal changes in twentieth-century physics—relativity theory for the large and very fast, and quantum mechanics for the very small—both relate to light. Everything is created from light, and all the interactions that follow after those created things are in place proceed by way of light. As for light itself, let us hear for a final time here that it stands outside the matrices of space, time, and matter that govern all of its creations.

If you suspect that I am leading up to saying that physics tells us that light is God, you are wrong, for I began this chapter by saying that science cannot touch that subject. But the boost that physics has given to light as a *metaphor* for God's creative activity is dazzling. If (and I emphasize the conditional here) God were to create a physical universe, what physicists describe sounds like how God might have gone about that job.

This section has approached light *objectively*, as a feature of the external world. Now we move to the way we *subjectively* experience light.

Light Subjectively Experienced

It stands to reason that if light symbolizes clarity, lucidity, and comprehension, darkness stands for their opposites. How could it be otherwise when in the dark we grope, stagger, stumble, and fall? Our disorientation erupts into our feelings. "Nobody feels good at four in the morning," a contemporary poem begins, and there is Gerald Manley Hopkins's memorable, "I wake and feel the fell of night, not day; / Self-yeast of the spirit, a dull dough sours."

This much is obvious. Now for something uncanny—as bizarre in its own sphere as are Einstein's discoveries concerning light. The story concerns a Frenchman, Jacques Lusseyran, as he tells it in his autobiography, *Let There Be Light.* It is not well known, so I will sketch its highlights.

Lusseyran's life must surely be numbered among the most remarkable on record. The age of nineteen found him serving as a vital link in the French resistance movement in the Nazi-occupied Paris of World War II, while at the same time preparing for the *Ecole Normale Supérieure* at the University of Paris. When a traitor betrayed the *Volontaires de la Liberté,* Lusseyran was arrested by the Gestapo and imprisoned for almost two years. When the United States Third Army arrived in April 1945, he was one of thirty survivors of a shipment of two thousand men who were on their way to Buchenwald. Stunned by their deliverance, Lusseyran tells us, the men could not at first even rejoice in their freedom.

With his freedom regained, however, something bizarre occurred. Despite the brilliance of both his academic and his wartime careers, Lusseyran was refused admission to the *Ecole Normale Supérieure* because of a decree passed by the Vichy government barring "invalids." For you see, Lusseyran had been totally blind since the age of eight. While roughhousing during a school recess, one of his friends had accidentally knocked Lusseyran over, slamming the back of his head into a sharp corner of the teacher's desk. The force of the blow drove a broken rim of his spectacles into his eyes, leaving him to live out his life in total darkness.

Or so we would suppose. I would not be telling this story here were it not for Lusseyran's report that in actuality something like the opposite occurred. "Being blind was not at all as I imagined it," he tells us. "Nor was it as the people around me seemed to think.

They told me that to be blind meant not to see. Yet how was I to believe them when I *saw?*"

Not at first, he admits. For a time he tried to use his eyes in their usual way and direct his attention outward, but then some instinct made him change course. In his own words,

> I began to look from an inner place to one further within, whereupon the universe redefined itself and peopled itself anew. I was aware of a radiance emanating from a place I knew nothing about, a place which might as well have been outside me as within. But radiance was there, or more precisely, light. I bathed in it as an element which blindness had suddenly brought much closer. I could feel light rising, spreading, resting on objects, giving them form, then leaving them. Or rather withdrawing or diminishing, for the opposite of light was never present. Without my eyes, light was much more stable than it had been with them.

Arthur Zajonc's *Catching the Light: The Entwined History of Mind and Life* provides a splendid, comprehensive survey of the subject of this chapter, but Lusseyran's account is the closest I have seen to a description of what Einstein's and Planck's light might feel like if we human beings could experience it directly. To what Lusseyran has already told us, I shall add only his report of the two virtues that invariably accompanied the light that reconstituted itself in him.

The first of these was joy. "I found light and joy at the same moment," he writes. "The light that shone in my head was like joy distilled, and from the time of my discovery, light and joy have never been separated in my experience." The connection was two-way. When negative emotions intruded on joy, his light became harsh, broken, jagged, and grating. In this sense, "fear, anger and impatience made me blind. The minute before I knew just where everything in the world was, but if I got angry, things got angrier than I. They mixed themselves up, turned turtle, muttered like crazy men and looked wild. I no longer knew where to put hand or foot, and everything hurt me."

The second virtue was that his intuitive powers were enhanced. "My sighted companions were nimble in bodily movements over which I hesitated. But as soon as it was a question of intangibles, it was their turn to hesitate longer than I."

It was this quantum leap in intuitive judgment that catapulted Lusseyran to leadership in the resistance movement. His ability to size up people's character and see through their dissembling was so uncannily accurate that he was put in charge of the delicate and dangerous job of recruiting, and everyone who applied to join the underground was sent to him for acceptance or rejection. His decisions were infallible, or (as he confesses) nearly so, for there was one man he admitted to the movement of whom he was not absolutely sure, and it was he who later betrayed them.

Lusseyran protested his being debarred from the *Ecole Normale Supérieure* on grounds of his blindness and he was admitted. After graduating with honors, he taught in France and in the United States at Hollins College, Case Western Reserve University, and the University of Hawaii. He was tragically killed in an automobile accident in 1971.

Conclusion

I close this chapter as I began it: God cannot be proved. There are sermons in science that beggar those in stones, but not *proofs,* and the same holds for phenomenological reports such as Lusseyran's. But the power in these sermons is remarkable. Because color with all its beauty screens the Pure Light of the Void, Goethe called it "light's suffering," which too is arresting.

What Christian (or at least Christian with the slightest metaphysical ear) can recite the Nicene Creed—"God from God, Light from Light, very God from very God"—without new comprehension after reflecting on the things that this chapter has touched on? What Jew or Christian can read God's first pronouncement, "Let there be light," without similar gain? It is not only Muslims who are moved by Rumi's lovely line, "Wist thou not that the sun thou seest is but the reflection of the Sun behind the veil?"

For myself, I add to the above what Reuben Snake once told me: "When we Indians first step out-of-doors in the morning, we raise both arms to greet the rising sun. And in an eruption of praise and gratitude, we shout 'Ho!'"

CHAPTER 9

Is Light Increasing? Two Scenarios

"Prediction," Niels Bohr told us, "is very difficult, especially about the future." When I first came upon his remark I thought it was a quip, but then I realized that he was probably separating predictions about the future from those about the outcomes of laboratory experiments. In any case, his point is incontestable: predicting the future is difficult to the point of being foolhardy. Even limiting prediction to the concern of this book—the human spirit—helps very little. Still, to "look before and aft, and pine for what is not," is built into the human makeup, so we have no choice. The first part of this book looked aft; this part looks ahead. Accuracy is in the lap of the gods, notably Chronos. Time will tell.

Taking my cue from weather reports, I lay out in this chapter two conflicting reports that we pick up from different stations. The first tells us that the skies are clearing after a major storm and the future of religion looks bright, even assured. ("Atmosphere clear, visibility unlimited," is a report that is seldom heard from air-traffic control towers, but it is on the books.) Meanwhile, from another weather station we hear the opposite. A tornado is approaching that could level religion forever. Beginning with the storm warning, I shall spell out these two reports and then align them in the manner of binocular vision.

God Is Dead

In the sixteenth and seventeenth centuries the scientific method replaced revelation, its predecessor, as the royal road to knowledge. Conceptually, it spawned the scientific *worldview,* while its technology created the modern *world.* The citizens of these new physical and conceptual environments constitute a new human breed whose beliefs correspond to very little in the human heritage. As a consequence, religion—the carrier of the traditional heritage—has been marginalized, both intellectually and politically.

First, politically. Easier travel and mass migrations have introduced a new phenomenon in history: cultural pluralism. The result has been to displace religion from public life, for religion divides whereas politics works for common ground on which citizens can adjudicate their differences. Concurrently, religion has been marginalized intellectually. Science has no place for revelation as a source of knowledge, and as modernists tend to think with science on matters of truth, confidence in revelation has declined. Marx considered religion "the sob of an oppressed humanity," and Freud saw it as a symptom of immaturity. Children who cannot accept the limitations of their actual parents dream of a Father in heaven who is free of those limitations. Theism is wish fulfillment, a pandering to "the oldest, strongest and most urgent wishes of mankind," and religious experience ("the oceanic feeling") is regression into the womb.

For religion to be marginalized socially and forced to the wall intellectually is no small event. Some see it as substantial enough to warrant the pronouncement that God is dead. Sociologists compile statistics on the change, but for the intellectual historian two developments suffice. First, on the question of God's existence, the burden of proof has shifted to the theists; and as proofs of the supernatural are difficult in any case, the classic proofs for God's existence have pretty much collapsed. The second sign, more telling, has already been remarked. Whereas atheists and theists used to agree that God's existence is an important issue, now even that common ground has vanished. The tension between belief and disbelief has slackened. It leaves no mark on intellectuals now. We have witnessed a decline in the urgency of the debate.

Such has become the common lot of most intellectuals. Their number has given rise to a distinction between *secularization* and *secularism*. The word *secularization* is now typically used to refer to the cultural process by which the area of the sacred is progressively diminished, whereas *secularism* denotes the reasoned stand that favors that drift. It argues on grounds that are cognitive, moral, or both that the desacralizing of the world is a good thing.

How, in the face of these seemingly irrefutable signs of faith's decline, is it possible to argue that the future of religion is promising?

The Eyes of Faith

One of the interesting recent developments in physics has been the realization that the observer must be included in experiments at that field's micro-frontiers. It is not just that we cannot know where a particle is until we perform an experiment to locate it. The particle is (from our side) literally nowhere until (by collapsing its wave packet) an experiment gives it location. This highlights the active component in knowing. Cognition is not a passive act. If seeing is believing, it is equally the case that believing is seeing, for it brings to light things that would otherwise pass unnoticed. In the words of William Blake,

This life's dim windows of the soul
Distorts the heavens from pole to pole
And leads you to believe a lie
When you see with, not thro', the Eye.

How does this affect the question of religion's future? The death-of-God prediction of religion's demise was reported through eyes that register data that is available to everyone. Religion, though, sees through the eyes of faith, and in doing so sees a different world. Or better said, it sees the same world in a different light.

In this new light things look different in ways that are overwhelmingly convincing. Arguments are irrelevant here, as they are when a rope that was mistaken for a snake is recognized for what it actually is. The sacred world is the truer, more veridical world, in part because it *includes* the mundane world. And in the act of including it, it redeems that world by situating it in a context that is meaningful throughout. As the Zen master Haquin put it, "This ground on which I stand / Is the shining lotus land, / And this body is the body of the Buddha."

Seen through the eyes of faith, religion's future is secure. As long as there are human beings, there will be religion for the sufficient reason that the self is a theomorphic creature—one whose *morphe* (form) is *theos*—God encased within it. Having been created in the *imago Dei,* the image of God, all human beings have a God-shaped vacuum built into their hearts. Since nature abhors a vacuum, people keep trying to fill the one inside them. Searching for an image of the divine that will fit, they paw over various options as if they were pieces of a jigsaw puzzle, matching them successively to the gaping hole at the puzzle's center. (We are back to Chapter Two and the women pawing over mountains of lingerie. Calvin likened the human heart to an idol factory.) They keep doing this until the right "piece" is found. When it slips into place, life's jigsaw puzzle is solved.

How so? Because the sight of the picture that then emerges is so commanding that it swings attention from the self who is viewing the picture to the picture itself. This epiphany, with its attendant ego-reduction, is *salvation* in the West and *enlightenment* in the East. The divine self-forgetfulness it accomplishes amounts to graduating from the human condition, but the achievement in no way threatens the human future. Other generations await in the wings, eager to have a go at life's curriculum.

The gulf that separates this faith-oriented projection of religion's future from the worldly one that was described is vast; but we live in the *uni*-verse, so in some way we must try to bring the two together. If we are religiously "unmusical" and religious accounts leave us cold, the situation is univocal: the God-is-dead pronouncement tells the tale. Those who have religious sensibilities, however, have a problem. The religious forecast carries weight, but so does the secular one. This is where binocular vision enters. How does the future of religion look when we take into account both what social analysts tell us—the first scenario—and what the eyes of faith register?

To bring those two accounts together we need to look again at the historical developments of the twentieth century, this time with eyes alert for signs that carry religious weight.

Clearing the Ground

Those signs come to light when we see that none of the secular jigsaw pieces the twentieth century reached for filled the space at the center of the human heart. The two most important ones were Marxism in the East and Progress in the West. (Marxism believes in lowercase progress too, but it stresses an ideological program for achieving it. As for *East* and *West,* I use those words to refer to the ideologies that polarized the twentieth century politically.)

Go back to the point with which I began this chapter—the scientific method that unhorsed the traditional worldview. That method's ability to winnow true from false hypotheses yielded *proven knowledge,* and proven knowledge can snowball. The eighteenth century's Industrial Revolution established those points historically, for by applying the assured knowledge that was ballooning, industrial advances raised Europe's living standards dramatically. Together the Scientific and Industrial Revolutions produced a third, psychological incursion—the Revolution of Rising Expectations.

That third revolution included several heady dreams that coalesced to form the Enlightenment: (1) thanks to science's reliable way of knowing, *ignorance* would be sent packing; (2) the reliable knowledge of nature that science afforded would send *scarcity* packing; and (3) the scientific worldview would send *superstition* packing. The superstitions that the Enlightenment had in mind were primarily those of the church, and with the church's back against the wall, it seemed that mankind was ready to advance into the Age of Reason. This Reason spelled Progress, the hope that has powered the modern world.

As for Eastern Europe and subsequently China, its version of that hope was, for the twentieth century, Marxism. To set that hope in perspective, we need only go back to the Revolution of Rising Expectations that the Scientific and Industrial Revolutions gave rise to. Hegel cashed in on the forward-looking stance of those revolutions and fashioned from it a worldview. From the seeming fact that things *were* getting better and stood a good chance of continuing to do so, Hegel extrapolated backward to infer that they had *always* been improving. Progress is the name of the game for being written into the nature of things. (In Hegel's vocabulary, Being is the necessary unfoldment of its Idea in ever-increasing consciousness and freedom.) Support for this heady scenario was welcomed from every quarter, and Darwin emerged to supply it from science. Inspired by Hegel, he painted the natural history of life on earth in strokes that fitted perfectly into Hegel's version of an evolution that is cosmic in sweep.

So far so good. But when we come to human history, Darwin's engine for advancement—natural selection working on chance variations—chugs too slowly to explain. A principle was needed to account for advancement in centuries, not eons, and Marx supplied it with his theory of class struggle. "Just as Darwin discovered the law of development of organic nature, so Marx discovered the law of development in human history," Engels intoned by Marx's grave at Highgate Cemetery.

One last step was needed, and though Marx assumed it, Engels (with Lenin's assistance) articulated it explicitly. To inspire not just hope but conviction, Communism's happy ending—the classless society—needed to be guaranteed, and that required metaphysics. For science is never enough, not even natural and social science together. To inspire conviction, hope needs to be anchored in the very nature of things. So Hegel's cosmic vision was reaffirmed, but with an important change. Its inclusive and forward-looking features were as they should be, but its vocabulary needed to be converted from idealism to materialism. This had the double advantage of making the theory sound scientific while at the same time directing attention to the politico-economic scene—specifically, to the means of production as the place where the gears of history grind decisively.

This is the package that in the twentieth century persuaded the Eastern half of humankind—the world's largest nation (those jurisdictions that made up the USSR) and its most populous one (the People's Republic of China). With its "jigsaw piece" (Communism) placed beside the piece the West reached for (Progress), we can proceed to the point for which I introduced them. Neither filled the spiritual hollow in the human makeup.

To begin with the West, Progress has turned into something of a nightmare. The campaign against ignorance has expanded our knowledge of nature, but science cannot tell us what we should give our lives to. That is disappointing: it is discouraging to discover that not only are we no wiser (as distinct from being more knowledgeable) than our forebears were; we may be less wise for having neglected value questions while bringing nature to heel. That possibility is frightening, for our vastly increased power over nature calls for more wisdom in its use, not less. The Enlightenment's second hope, of eliminating poverty, must face the fact that more people are hungry today than ever before. As for the belief that the Age of Reason would make people sane, that reads today like a cruel joke. In the Nazi myth of a super-race (which produced the Holocaust) and the Marxist myth of a classless utopia (which produced the Stalinist Terror and Mao's Cultural Revolution), the twentieth century fell for the most monstrous superstitions the human mind has ever embraced.

With this last point we have already moved to the Eastern half of the twentieth century, where Marxist hopes have not just declined but collapsed. The Soviet Union is in shambles, and while Maoism remains nominally in place in China, no one believes it anymore—capitalism is advancing there faster than anywhere else on the planet. In its heyday, Marxism inspired commitment by claiming that its idealism was grounded in truth. This is indeed the winning formula, but the twentieth century falsified both halves of the Marxist vision. All of Marx's major predictions turned out to be wrong: (1) the European model of production has not spread throughout the world; (2) the working class has not progressively grown more miserable and radical; (3) nationalism and religious zeal have not declined; (4) Communism does not produce goods more efficiently than free enterprise or distribute them more equitably; and (5) in Communist countries, the state shows no signs whatever of withering away.

Faced with this miserable predictive record, apologists regularly switch from truth to justice—are we to forget the suffering masses? But the Marxist record on compassion is no better than its record on truth. In justifying Communism's (often demonic) means on grounds of the humanitarian ends they were supposed to lead to, Marx saddled his movement with a bloody-mindedness the likes of which history has rarely seen.

Modernity's coming to see the gods it worshiped for what they were—idols that failed—was the most important religious event of the twentieth century. With the ground cleared of those illusions, we can now inspect the scorched earth to see if it shows signs of new life.

CHAPTER 10

DISCERNING THE SIGNS OF THE TIMES

At least we can say that religion has weathered the storm. On his seventy-fifty birthday, Malcolm Muggeridge looked back over his long worldwatch as editor of the *Manchester Guardian* and concluded that the most important single *political* fact of the twentieth century had been that with every means of suppression at its command for seventy years, the USSR had not been able to destroy the Russian Orthodox Church.

I can add to Muggeridge's observation the survival of the Christian Church in China under similar circumstances. When my missionary parents left China in 1951 after nine months of house arrest under the Communists, they thought that their lifework had been in vain. Thirty years later, when I returned to visit my childhood haunts, the ban against organized religion had just been lifted, and the vitality that the church had maintained in its underground years surprised everybody. To make sure I could locate the large church we used to attend when we were passing through Shanghai, I reached it forty minutes before the Sunday morning service was to begin and found standing room only. Sixteen Sunday school rooms that were wired for sound were likewise packed, and during the announcement period in the service the pastor pleaded with the congregation not to attend church more than once each

Sunday, for that deprived others of the opportunity. (It has been a while since I have heard that plea in my church.) After the service, while lunching with the retired pastor of the church (who had learned his English from my father), I heard firsthand stories of what Christians had had to endure during the Cultural Revolution—being forced to wear dunce caps and kneel for two hours on broken glass in front of jeering mobs, and the like. Those stories concerned Christianity, the "foreigners' religion," but Muslims and Buddhists suffered too and have made comparable comebacks. Mao excoriated Confucius as being bourgeois, but Confucian ethics is back in the schools again.

Such resilience has brought even those who are not themselves believers to respectful recognition of religion's durability. Having discovered no society without religion, anthropologists (riding their functionalism, which holds that institutions that serve no purpose fall by the wayside) now consider religion adaptive. Neuroscientists trace its utility to the very structure of the human brain; once the left-brain interpreter was fully in place and reflexively active in seeking consistency and understanding, religious beliefs were inevitable, we are told. Having authored *The Joy of Sex,* the gerontologist Alex Comfort stands in no danger of being charged with excessive piousness, but his verdict is similar to the one just given: "Religious behaviors are an important integrator of man's whole self-view in relation to the world." Carl Jung reached his conclusion, which he stated categorically, from his analytic practice: "Human beings have an in-built religious need." Philip Rieff, a leading authority on Freud, draws this paragraph together by likening faith to the glue that holds communities together, adding that the weakening of this glue in the twentieth century has changed Dostoyevsky's question, "Can civilized men believe?" to "Can unbelieving men be civilized?" André Malraux's most celebrated dictum is to the effect that the twenty-first century will be religious or there will be none.

These quotations are clear indication that thinkers are again taking religion seriously. That, though, does not touch the question of religion's truth. Informed thinkers now believe in *religion;* do they believe in *God?* Some do and some do not, of course. What follows is an assessment of the general scene, with special attention to changes that seem to be afoot.

Straws in the Wind

Several years ago the *New York Review of Books* noted that "a revival of theism seems to be taking place among intellectuals." One important bit of evidence of this trend is the founding of the Society of Christian Philosophers. In mentioning that society earlier, I noted that the philosophy of religion (to say nothing of Christian philosophy) is not in good standing in the eyes of the philosophical establishment; still, the very emergence of such a society in the last quarter-century suggests a change. Including in its membership over sixteen hundred of the ten thousand members in its umbrella organization, the American Philosophical Association, most of them younger members, it publishes a first-rate journal, *Faith and Philosophy,* which carries Tertullian's dictum, "faith seeking understanding," on its masthead. Even philosophers such as Levinas, Heidegger, and Derrida (who not only are not Christian but resist the epithet *theist*) find their latest writings sounding surprisingly like the "negative theology" of mystics, whose God is hidden in the "cloud of unknowing."

Philosophy occupies a special place in a book on worldviews, which is why I gave it this initial glance. Having done that, I now turn to the broader scene. To get started, I shall hold my finger to the wind to try to get an impressionistic sense of the direction in which it is blowing.

- *Item.* In titling his influential book *A History of the Warfare Between Science and Religion,* Andrew Dickson White continued the late-nineteenth-century view that science and religion were locked in a battle that science was certain to win. Intellectuals adhered to that view for most of the twentieth century; in 1965 the historian Bruce Mazlish (then my colleague in the humanities department at MIT) wrote that it was "beyond reasonable doubt that the warfare model continues to remain solidly entrenched as the dominant one." That has changed. Scientific triumphalism has peaked, and hope increases for a peaceful coexistence.
- *Item.* A breakthrough book in the academic study of religion has appeared. Contrasting sharply with the antipathy toward religion that social scientists showed during the nineteenth and twentieth centuries, at the twentieth century's close Roy Rappaport (onetime president of the American Anthropological Association) argued in his *Ritual and Religion in the Making of Humanity* that religion has been central to evolution since the human species appeared, and will continue to be central to any cultural advance we may achieve from this time forward.
- *Item.* Chancellor David K. Scott of the University of Massachusetts, Amherst—a physicist by training—foresees a return of religion to the university, not just in departments of religious studies (leaving other departments free to ignore the subject), but in ways that would enable students to wrestle throughout their college years with the issue of ultimate values, meanings, and purposes. This is not a return to the medieval university, he says; it merely faces up to the fact that the pendulum has swung too far in the opposite direction.

Scott envisions an "integrative university" in which spirituality will be an ally rather than an enemy in educating students to be engaged citizens in an enlightened democracy. The Constitution is no obstacle, he thinks. The university has been using it as a "convenient barrier" against taking religion seriously.

- *Item.* Sales of religious books have risen spectacularly (by 50 percent in the last ten years), and religion is entering into the deep grain of some our most respected writers. Flannery O'Connor and Walker Percy carried the torch through the lean years following T. S. Eliot, Graham Greene, and W. H. Auden; and Saul Bellow, Tom Wolfe, and John Updike are only three notable writers who are sensitive to thin places on the line that separates this world from another. Updike characterized Tom Wolfe's *A Man in Full* as "all about religion." Unlike the other two writers named, Updike is explicit about his religious views: "If this physical world is all, then it is a closed hell in which we are confined like prisoners in chains, condemned to watch the other prisoners being slain."
- *Item.* The shibboleth of "sticking to the facts" requires journalists to distance themselves from what they report, but it is not lost on them that religion now sells. Every journalist worth his or her salt, Bill Moyers has said, knows that the towering question of our time is, What is the human spirit? And the question is being addressed on many fronts. *Newsweek*'s 1998 cover story (which carried the interesting-if-true banner "Science Finds God") was quickly emulated by its competitors. *Business Week* devoted the cover story of its 1999 seventieth-anniversary edition to "Religion in the Workplace: The Growing Presence of Spirituality in Corporate America."
- *Item.* One of the surest signs that something has entered the public consciousness is cartoonists' picking it up. A few read-

ers will remember the *New Yorker* cartoon that pictured an executive sitting cross-legged on his desk while his secretary staved off an arriving visitor by telling him, "I'm sorry, but Mr. Mason is at one with the altogether at the moment." Televangelists now take second place to high-audience-rated serials such as *Touched by an Angel;* and more than fifty years after it first went on the air in 1950, *Unshackled,* a born-again Christian radio program, is still going strong, carried by 1,200 stations in over 140 countries.

- *Item.* Secular humanism is no longer the confident battle cry that it was when, in 1933, intellectuals gathered around John Dewey to hammer out the Humanist Manifesto. That initial manifesto was solidly mainstream. Beginning with Dewey himself, its signatories included recognized names in every field of culture—Isaac Asimov, John Ciardi, and B. F. Skinner among them. The second, updated manifesto (1973) already sounded more defensive than confident, and the third (1999) comes close to reading (dully) like a shutdown. Not a single recognized name appears among its signers.
- *Item.* The resources of spirituality for mental health are being seriously explored. Five times in the last three months, from Seattle to the University of Florida, I have been asked to address audiences consisting exclusively of psychologists, psychiatrists, and psychiatric social workers on the interface between their work and mine. A network of groups such as those that invited me seems to be gathering strength.

How many robins does it take to make a spring? The above list could be extended, but a list of counterexamples, equally long, could easily be compiled to place alongside it. Readers would have to decide for themselves which list they found more telling.

Counterculture and the New Age Movement

I think it was Nathan Pusey who characterized Harvard University as an assemblage of autonomous departments united by a central heating plant, and it is a telling quip. Communities of scholars are a thing of the past, but I have witnessed one exception. During two of my MIT years the miracle happened. Chaplains at the Institute joined forces and mounted a series of monthly meetings on "Technology and Culture" that *worked.* Each month a prominent member of the faculty was invited to air personal views on a social issue that concerned him or her, and a general discussion followed. We filled the auditorium each time and experienced our Institute in a new way.

I mention this for the one program in the series that I remember. I cannot recall who the speaker was, but I clearly recall how he began his presentation. Holding aloft Theodore Roszak's *The Making of a Counter Culture* (which had recently been published), he told us how much it had upset him. He had read it through twice, he said, trying to understand why young people would be hostile to the science that was the love of his life.

I pricked up my ears, for I had just written an article (rather archly titled "Tao Now") in which I argued that the Asian and Western views of nature had collided in a single phoneme. "Tao" and "Dow" are pronounced identically, and we were at the height of the Vietnam protest, which had turned the Dow Chemical Company into a symbol for the Pentagon through its manufacture of napalm. Which of the semantic options in that phoneme, I asked—one profoundly ecological, the other violently destructive—did we want to commit the human future to?

Needless to say, the speaker of the afternoon came down hard on the distinction between science and the uses made of it, but (I speak for myself) the underlying fact in the situation will not go away. Because (1) science has given us unforeseen power over nature, and (2) human beings do not have the wisdom and virtue to keep from using that power for private gains that work against the common good, science is no longer seen as the messiah that will save us.

Roszak's counterculture was enraged primarily by the destructive uses of technology, whereas its successor, the New Age Movement, picks up the other side of the science story: its worldview and the strictures it places on our full humanity. The advocates of that latter counterculture want out—out from the prison of that outlook. Because they lack seasoned guides, their unbridled enthusiasm for the Aquarian Age careens crazily, and conceptually the movement is pretty much a mess. Pyramids, pendulums, astrology, ecology, vegetarianism and veganism (we are back to religion as dietary restrictions); amulets, alternative medicine, psychedelics, extraterrestrials, near-death experiences, the archaic revival, channeling, neopaganism, and shamanism—these and other enthusiasms jostle one another promiscuously. And brooding over them all is Gaia—Gaia and the goddesses (both within and without). Flaky at the fringes and credulous to the point of gullibility—an open mind is salutary, but one whose hinge is off?—the New Age Movement is so problematic that I would gladly leave it alone were it not for the fact that it has two things exactly right. First, it is optimistic, and we need all the hope we can get. Second, it adamantly refuses to acquiesce to the scientistic worldview. Instinctively it *know*s that the human spirit is too large to accept a cage for its home.

Four Modern Giants Revisited

After Kant and Hegel, the principle architects of the modern mind have been Darwin, Marx, Nietzsche, Freud, and Einstein. Einstein has not threatened religion. Asked if he believed in God, he answered, "Yes, Spinoza's God"; and while that is not the God of Abraham, Isaac, and Jacob, it qualifies. The other four, however, have given religion a hard time. Reduced to slogans, their respective charges—religion is opiate, illusion, slave-mentality, and surplus baggage—have echoed like clichés in the modern mind. Together, their authors constitute a formidable phalanx for the human spirit to confront. It is a significant sign of our times, therefore, that their theories are being challenged. Not everything in their theories; all four will retain their places in history. But they will be remembered with more reservations than were initially in the picture.

Charles Darwin

That life on this planet began in a primal ooze and (over the course of some three and a half billion years) has advanced to human complexity—these points seem solidly in place. As does Darwin's discovery that natural selection working on chance variations plays a part in the process. But even when chance and other observable causes (such as historical contingencies and changing environments) are added to the picture, natural selection is not turning out to have as much explanatory power its author expected that it would. Taken as a complete explanation of human origins, his theory is beginning to resemble the Ptolemaic hypothesis on its last legs: each time a difficulty is encountered, another epicycle is added to salvage the theory. Are transitional forms conspicuously lacking at certain points in the fossil record? Punctuated equilibrium comes forward to account for the gap: it happened too fast (only thirty million years or so) to leave enough deposits to be noticed. The debates rage like wildfire, but (to all but those who have made up their minds) one point emerges clearly: Darwinism has not edged God out of the evolutionary process as Darwin thought it would.

I continue anecdotally. In the summer of 1997 I was invited to give a week-long series of lectures at the Chautauqua Institution in upstate New York. For one of the lectures I chose the topic of evolution, and in the course of my comments I read an open letter to the National Association of Biology Teachers (NABT) that I had composed. That letter asked if the association might consider dropping two inflammatory words from its official definition of evolution, which definition I then quoted: "Evolution [is] an *unsupervised, impersonal* natural process of temporal descent with genetic modification that is affected by natural selection, chance, historical contingencies and changing environments" (italics added). Had biologists discovered any *facts,* my letter went on to ask, that prove that the process is "unsupervised" and "impersonal"? If not, would the association consider deleting those two words, which many Americans see as threatening their belief that God had a hand in the process?

Seizing the opportunity to vent the ham in me that I share with all born teachers, I had brought with me a stamped, addressed envelope. Folding the letter with a theatrical flourish, I stuffed it into the envelope, sealed the flap, and (at the lecture's close) walked ostentatiously to the Chautauqua mailbox, where I posted the letter with a resounding clank of the lid.

Ten days later, back at home, I received a reply from the executive director of the NABT. It began by thanking me for the civil tone of my letter. Most of the correspondence the association receives, the director said, tells them that they are agents of the devil and going to hell in a wheelbarrow, but my letter was at least courteous. With that, he went on to note that the association's board would be meeting later that month and that he intended to place my letter on its agenda.

I found this interesting and alerted the Religious News Service to what was up. The story it released at the close of the episode contained these facts.

The board of the National Association of Biology Teachers meets for four days once a year. On the first day of that year's meeting, its members considered my letter for ten minutes and voted unanimously to reject its suggestion. The issue refused to die, however. In elevators and corridors, during cocktail hours and in individual encounters throughout the week, members of the board continued to discuss it, and as the final item of that year's meeting they brought the letter back to the table. This time they discussed it for forty minutes, after which they voted unanimously to reverse their previous decision.

So those two words are gone from the official NABT definition of evolution—but the story does not end there. Remember, I am using this anecdote to indicate what in Darwinism needs to be left behind, and my tilt with the windmill entered a second round two years later.

Flushed with my initial success, when the Kansas evolution brouhaha erupted in 1999, I decided to press my luck and sent a second suggestion to the NABT. This time I proposed that its board consider recommending that, at the first class meeting of every course dealing with evolution, teachers distribute a handout that would read something like this:

> This is a course in science, and as your instructor it is my responsibility to teach you what science has empirically discovered about the mechanisms by which life emerged and has developed on this planet. We scientists are convinced that we know an important part of that story, and I will do my best to inform you of it.
>
> However, there is so much that we still do not know that

plenty of room remains for you to fill in the gaps with your own philosophic or religious convictions.

I added that the phrase in my letter, "There is so much we do not know" was a quotation from Stephen Jay Gould.

My letter was acknowledged by the executive director, but with the notation that he doubted that this second proposal would be accepted. That was the last I heard of the matter.

Looking back on this second try, I freely grant that the tactic I proposed may not have been the right one, but I fail to see why an olive branch of this sort would not (without harming science in any way) do us all good. Evolution will come up again in the next chapter, but for here I have made my point. As (to date) a partial explanation for how we got here, Darwinism should be taught, and efforts to fill in the gaps in the theory should continue. But claims to the effect that Darwinism is so much on top of the story that it is unreasonable to think that other causes (some of which might not be empirical) could have played a part—that proscription should be dropped.

Karl Marx

Marx's compassion for the downtrodden, especially as it comes through in his early writings, should never be forgotten, for it rivals the power of the Hebrew prophets to move and motivate. Forty years in the British Museum to hammer out a reading of history that would hold hope for the masses is dedication of a magnitude that rivals Mother Teresa's. The soundness of his program for realizing his vision is another matter.

The preceding chapter has already pointed out that none of Marx's predictions came true, but I am certainly in no position to fault him for that, given that my book too plays with the future.

What must be faulted is Marx's deep mistake in placing his faith in social engineering. If the oppressed masses were "sobbing" (Marx's moving word) before the Russian revolution, they were writhing in desperation in the postrevolutionary gulag. Whether the Soviet death toll was one hundred million *of its own people* (which is probable) or a "mere" ten million (which is implausible), the eighty years of terror is unforgivable. It will be objected that Marx's vision was betrayed, and that is true respecting specifics. But it is not specifics that I am speaking of. It is Marx's faith in social engineering, and his willingness to sacrifice whatever was necessary for that faith.

Those who knew Karl Popper tell us that his dictatorial ways did not offer a good model for the society he argued for in *The Open Society and Its Enemies,* but we all preach better than we practice, so that should not be held against the truth of the book's thesis. Attempts to revolutionize the world to make it conform to a preconceived ideology can lead only to totalitarianism, for history is too unmanageable to conform to our conceptual designs. Instead of attempting "holistic engineering" (Popper's phrase), we should approach things gradually—cautiously and incrementally—as we keep close watch on the signals our initiatives send back to us. Pull out the stops on policies that prove effective, while being wary of formulas that assume that we understand how history works. Above all, do not inflict real suffering for the sake of hypothetical panaceas. Ends do not justify means, as Marx mostly thought they did.

Marx was right, though, in documenting the extent to which class interests skew the ways people see the world. (He would not have been surprised by the finding that when poor children are asked to draw a nickel, they draw it larger than do children of the well-to-do. It looms larger in their world.) Reinhold Niebuhr was strongly influenced by Marx in this, and in his classic *Moral Man and Immoral Society* (which at one point John F. Kennedy and Che

Guevara were reading concurrently) documented the outworkings of this point in twentieth-century America. Niebuhr winnowed Marx's thinking accurately, seeing him as right about class interests but wrong in advocating social engineering.

On the latter point, Niebuhr's book preceded Popper's (though there is no evidence that Popper read Niebuhr) and gave the issue of social engineering a theological overlay that Popper's did not. To think that we can circumvent the human heart and, in the face of its unregenerated clamorings, achieve the kingdom of heaven by revamping social institutions is to overlook the fact that that kingdom is first and foremost an interior affair. Short of the *eschaton* (end of history), its arrival in the world will be in proportion to its arrival in human hearts.

This does not counter the importance of social action. (Niebuhr was an all-out activist. To oppose the Communist-controlled "united front" after World War II, he joined with Eleanor Roosevelt to found Americans for Democratic Action.) But it does insist that if such action is to be productive, it must proceed in the understanding that "there never was a war that was not inward" (Marianne Moore).

What about religion as the opiate of the people? That condemnation is a useful reminder of a danger that religion can (and repeatedly does) succumb to, but it is a half-truth. Religion has, on occasions, been a conservative force, and on others it has been revolutionary in intent and achievement. It has been an opiate, and also a stimulant. It has identified too closely with particular cultures, and it has challenged the status quo. It preoccupies itself with raising church budgets, and with raising the oppressed. It makes peace with iniquity, and tries to redeem the world.

Finally, there is the adage that Marx found Hegel standing on his head and stood him on his feet, which translates into the fact that Marx accepted Hegel's historical determinism while replacing his idealism with materialism. Historical determinism has gone by the board, but materialism remains a philosophical option. Those who are drawn to it are entitled to their convictions, but it has no lien on those who find it wanting.

Friedrich Nietzsche

One day, during the time we were studying Nietzsche in a history of philosophy course I was teaching at Washington University, an exceptionally involved and vocal student did not show up. When his absence continued to the point of being noticed, one of his friends looked him up in his dormitory room. He found him lying in bed reading *Thus Spake Zarathustra.* Asked whether he was ill, he said no, and when the natural question followed, "Why aren't you coming to class?" he answered, "You guys are still looking for it. I've found it."

Many people will resonate to that answer, for there are few who have not at one time or another fallen under Nietzsche's hypnotic spell. He electrifies. But what are we to say about the truth of his words?

The close of this student's story does not provide an encouraging answer. A large, athletic man, he took Nietzsche's doctrines of the superman and the will to power to include physical strength. Intoxicated by the notion, he egged his dormitory rival into a fight, which he lost catastrophically. His opponent mopped the floor with him. Disgraced, he left the university, and that was the last we saw or heard of him.

If this was a misreading of Nietzsche, Nietzsche left himself open to it. It would be wrong to lay the blame for Nazism directly at his door, however—many things contributed to that madness. In any case, it is a different contention of Nietzsche's that is of direct concern here—namely, his depiction of Christianity. Here, as is so often the case, Nietzsche gave us half-truths. Resentment and slave mentality have figured in Christian history, but so have courage and compassion. Are we to consider Martin Luther King, Jr., a slave? Or Dietrich Bonhoeffer, whom the Nazis imprisoned and then executed for helping the Jews? Such men are symbols for the millions of their like who risked their lives equally.

Bonhoeffer merits a further word. As professors have been the prime purveyors of Nietzsche's notion that Christianity is an embodiment of slave mentality, it is worth pointing out that German churches resisted Nazism (not as much as history would like them to have done, but it is easy to condemn from a safe distance), whereas the universities did not. If Bonhoeffer symbolizes the first camp, Heidegger tokens the second.

As for Nietzsche's madman charging through the streets crying, "God is Dead," that tale stands (as we saw in Chapter Three) with Plato's allegory of the cave as the second of the two great allegories that frame Western civilization like majestic bookends. Nietzsche's prescience was inspired; it is not too much for me to say that the tunnel in the first half of this book of mine does nothing more than detail the message that Nietzsche's madman tried to get across to his heedless bystanders.

What has recently become clear (through the publication of some revealing conversations between Nietzsche and several of his closest friends) is how Nietzsche himself reacted to the news of God's death. The following short account by Ida Overbeck in Sander Gilman and David Parent's *Conversations with Nietzsche* says it best:

> I had told Nietzsche that the Christian religion could not give me solace and fulfillment. I dared to say it: the idea of God contained too little reality for me. Deeply moved, he answered:

> "You are saying this only to come to my aid; never give up this idea! You have it unconsciously. One great thought dominates your life: the idea of God." He swallowed painfully. His features were completely contorted with emotion, until they then took on a stony calm. "I have given him up, I want to make something new, I will not and must not go back. I will perish from my passions, they will cast me back and forth; I am constantly falling apart." These are his own words from the fall of 1882.

For as long into the future as we can see, critics will continue to argue over what Nietzsche really intended. His greatness, though, will not diminish, if for no other reason than (as William Gass puts it) "he bit our values as if they were suspicious coins and left in each of them the indentation of his teeth."

Sigmund Freud

My wife recently reminded me of something I had forgotten. At a dinner for Aldous Huxley while he was paying a campus visit to Washington University, I asked him if there were any books he found himself returning to reread. Actually, he said, there were two. One was Sir Herbert Read's *Art and Education*, and the other was a book no one at the table had heard of, Ian Suttie's *The Origins of Love and Hate.* A psychologist herself, Kendra followed up on the second book, and this is what she found.

Its author, a Scottish psychiatrist, is something of a mystery. He died in midcareer in the 1930s, and his work was ignored until a student of his, John Bowlby (who became a recognized name in child development) took it up and brought out a second edition of his neglected book.

Like Freud and other psychologists, Suttie believed that people deal with anxiety by shoving worrisome thoughts and feelings into the unconscious. But unlike psychoanalysts, he became convinced (through his research) that our major repression is not of sexual or aggressive impulses, but of affection and openness. These repressions in individuals add up to a collective taboo against tenderness in our culture.

Beginning at the beginning, Suttie saw infants as born with two independent propensities. The one that is primary is a desire for the social give-and-take and responsive relationship that we call *love.* Sexuality, in his theory, exists as a separate and independent drive.

This is a radical departure from Freud, who likewise posited two independent drives, one of which was sex (the libido) and the other aggression (the death instinct). Freud described the infant's earliest state of consciousness as auto-erotic and narcissistic. In contrast, Suttie describes the earliest state (before an infant distinguishes self from others) as a state of symbiotic communion.

In the Freudian view the infant believes itself to be omnipotent, able to summon the mother magically with its cries. It cathects to the mother because she relieves its bodily tensions. To Suttie this was as preposterous as saying that the mother loves the baby because it is a breast-reliever who drains her swollen mammary glands.

In his years of careful scrutiny, Suttie became impressed by the early overtures that a baby makes to evoke a response from its mother. It fixes its gaze raptly on her face while nursing, and this often gets a loving gaze in return. Soon the baby starts smiling around the nipple as it sucks, often dropping the nipple and gurgling with delight if it gets a response. It is a mode of flirtation. In the harmonious interchange between mother and child, the baby gives the only thing it can, its love and its body as the first shared plaything. It is the beginning of the creativity that Suttie sees in play, which (he says) is the mother of invention, not necessity.

A critical period comes when the infant is able to differentiate itself from its mother, and its mother from other persons. It is only then that the baby can know separation from its mother, and this separation is the major source of human anxiety—a fear of abandonment. At about the same time acceptance is no longer unconditional. Some of the baby's bodily functions and activities may not be welcomed and approved.

In baby or adult, hell hath no fury like rejected love. Here we have Suttie's understanding of the origin of anger, which he saw as a baby's desperate effort to reclaim a lost harmony. Depending on the degree of pain and hopelessness the small child goes through, intimacy may be renounced, and a quest for self-sufficiency (or power) may take its place—the typical route in our individualistic West, Suttie believed.

Peter Koestenbaum, who conducts seminars for industrial leaders, offers this bit of documentation. In one of his discussion groups he paused and said, "Sometimes it is necessary to speak from the heart." "The heart is a pump," an oil tycoon grumped. Later, when Koestenbaum had softened some of the man's defenses, the tycoon showed him a secret. Locked in a drawer of his desk was heartfelt poetry that the man had written. He said that he had never shown it to anyone before, fearing that in doing so he would lose his authority.

It is as Suttie says: tenderness is a cultural repression. Suttie cites obsessive, compulsive sex as another possible outcome of repressed tenderness, for (as the saying goes) you cannot get enough of what you do not really want. What is needed and wanted (but our culture denies, Suttie argues) is emotional closeness. Not sex, nor food, nor power, nor any other surrogate can satisfy that need.

I have almost exhausted my allotted space for Freud and I have scarcely mentioned his theories. P. B. Medawar may have gone too far in calling Freudianism the greatest hoax of the twentieth century, but if Adolph Grunbaum and Frederick Crews have not shown how little there is by way of reason and hard evidence to require us to accept his self-admittedly loveless view of human nature, I am not going to accomplish that task here. The American Psychoanalytic Association is always ready in the wings to come back with its charge that the critics just don't get it, though its tone sounds more defensive every decade.

What we need is an alternative. My strategy—to present an alternative to Freud's theories rather than argue against them—occurred to me through remembering one of my all-time favorite sermons. In the nineteenth century, sermons had robust titles, and this one (by Thomas Chalmers) was titled "The Expulsive Power of a New Affection." The idea for it came to him, he tells us, while he was in a coach being driven over a mountain road. At a narrow place where the road was at the edge of a steep precipice the coachman began to whip his horses which seemed to Chalmers a dangerous procedure, but the driver said he must divert their attention from the danger of the road. The sting of his whip gave them something different to think about.

It is no different with human beings, Chalmer's sermon went on to explain. People do not shake off familiar habits by dint of reason and willpower. They need something new to think about and respond to. I hope that Ian Suttie provides the sting of the whip that is the needed stimulus here.

CHAPTER 11

THREE SCIENCES AND THE ROAD AHEAD

The scientific worldview presents itself as a stupendous story. I sketched its outline in the second chapter of this book, but on closer examination it looks like three novellas strung together by sequential questions: How did the universe get here? How did life get here? How did we human beings get here? Physics addresses the first question, biology the second, and biology plus the cognitive sciences the third. These are the three sciences that have the largest metaphysical implications, and if they are moving toward a less-entunneled outlook, that is the best sign we could have that the outdoors is drawing closer.

PHYSICS

It is starting to look as if physics is out of the tunnel already. I say that on the authority of the EPR (Einstein-Podolsky-Rosen) experiment, which establishes that the universe is nonlocal. Separated parts of it—how widely they are separated makes no difference; it could be from here to the rim of the universe—are simultaneously in touch with one another. In lay language, what the EPR experiment demonstrates is that if you separate two interacting particles and give one of them a downspin, instantly the other will spin upward.

The theoretical consequences of this finding are revolutionary—sufficiently so for Henry Stapp of the University of California, Berkeley, to call it "*the* most important finding of science, ever," for it relegates space, time, and matter (the matrices of the world we normally know) to provisional status. If we were to look out upon the world through a window with (say) nine panes of glass set in place by latticework, we would *see* the outdoors as divided by the latticework (which of course is not in the landscape we are looking at). Something like that pertains here.

What are the implications of all this? Let us take a look.

Everything we perceive with our senses (and analyze and classify into laws and relationships) has to do with the relative world, a kind of phantom play of names and forces flowing temporally in the stream of space and time. In this relative world there are no absolutes; time and change govern everything. Nowhere are there fixed frames of reference, nowhere objects that can be considered independent of their observing subjects. No event can be perceived in exactly the same way by all observers, and there is an irreducible uncertainty that precludes the possibility of our ever knowing all the fundamental properties of the phenomena that we experience and investigate. This uncertainty is built into the very fabric of the universe, so nothing escapes it. The whole cannot be reduced to a set of basic building blocks, for on the cosmic scale matter can disappear into pure energy and reappear in a different guise. The ancients would not have been surprised. *Anicca, anicca;* impermanence, impermanence. Maya reigns, and Shiva's tireless dance continues.

But that is only half the picture. What puts post-EPR physics all but outside the metaphorical tunnel this book revolves around can now be stated explicitly. The moment of truth in the EPR experiment opens a rift in the cloud of unknowing through which physicists catch sight of another world, or at least another reality. Again

I will let Henry Stapp say it: "Everything we [now] know about Nature is in accord with the idea that the fundamental process of Nature lies outside space-time, but generates events that can be located in space-time." Stapp does not mention matter, but his phrase "space-time" implies it, for physics locks the three together. Stapp's close associate, Geoffrey Chew, dubbed that word in explicitly when (during a dinner party at which I seized the opportunity to sit next to him) he said to me, "If you begin with matter as a given, you're lost."

New Age enthusiasts are quick to jump in here with the announcement that physicists have discovered God, which of course is not the case. All physicists have found is that what runs the show (runs the spatio-temporal-material universe) lies outside that show. Still, in establishing the existence of "something," if only a not-further-characterized X, beyond the spatio-temporal-material world, nonlocality provides us with the first level platform since modern science arose on which scientists and theologians can continue their discussions. For God too resides outside those three perimeters.

There are some who will think that if I stop with nonlocality and do not add to it Intelligent Design I will be overlooking an important second reason for saying that physics is out of the tunnel, so I shall speak to it, though in the end I will not bank on it.

More and more, scientists are finding that if the mathematical ratios in nature had been the slightest bit different, life could not have evolved. Were the force of gravity the tiniest bit stronger, all stars would be blue giants, while if it were slightly weaker, all would be red dwarfs, neither of which come close to being habitable. Or again, had the earth spun in an orbit 5 percent closer to the sun, it would have experienced a runaway greenhouse effect, creating unbearable surface temperatures and evaporating the oceans; while on the other hand, if it had been positioned just 1 percent farther out, it would have experienced runaway glaciation that locked earth's water into permanent ice. On and on. We get the point.

Physicists of the stature of John Polkinghorne find it impossible to believe that such fine-tuning (and the apparent frequency with which it occurs) could have resulted from chance. They toss around improbability figures in the range of one in ten followed by forty zeros. For them, improbabilities of this order all but require us to think that the universe was designed to make human life possible, to which they add that design implies an intelligent, intentional designer. They do not laugh when a fellow scientist, Dale Kohler, writes, "We have been scraping away at physical reality all these centuries, and now the layer of the remaining little that we don't understand is so thin that God's face is staring right out at us."

I am not myself a scientist, but I naturally favor the Design hypothesis. At sea with numbers higher than "the ten thousand things" (the archaic Chinese phrase for heaven and earth, the universe), ten followed by forty zeros completely escapes me. Still, a single fact can carry me to the conclusion the ratios I cited suggest. If the Andromeda Galaxy were not there, neither would we be—we are, quite literally, made of stardust. This is quite enough to blast me into a moment of mystical frisson.

However, that is me—me in the company of ranking scientists who share my spiritual sensibilities. The problem with citing a must-have-been-designed universe as an added indication that physics is out of the tunnel is that an equal number of qualified physicists—Stephen Hawking, for one—disagree with this reading of the matter. Whether the disagreement turns on evidence or on the philosophical lens through which the evidence is viewed is itself at the heart of the controversy. Because the evidence is beyond my competence to weigh, any call I made in the dispute would reflect nothing more than my own beliefs and perceptions and thus would count for nothing. It is a good sign that the issue is being vigorously discussed, and no one can fault believers for finding in Intelligent Design a resource for their faith. But that is the most that can be said at this point in the dispute.

While I am backing off, I should go back to nonlocality and admit that physicists disagree over its implications too. When I asked Geoffrey Chew whether he and Stapp were mainstream physicists or mavericks, he answered cheerfully, "Oh, we're mavericks all right, but our number is increasing every year." What leads me to declare that physics is all but out of the tunnel is my faith that the increase will continue.

BIOLOGY

I shall not be dealing with molecular biology here—DNA and the three billion chemical letters of which human genes are made. The technological possibilities of bioengineering (for good and for ill) are gargantuan, but their metaphysical implications are modest, so I shall pass over those aspects of biology and go straight for the prize, Charles Darwin.

Writing a book under a deadline renders one illiterate for the interval, so David Walsh's book on *The Third Millennium* had passed me by until my eye fell on an ad for it as I was sitting down to write this chapter. Once again (as in the chapter on law), I can only credit Providence with the timing, be that superstition or not. Walsh's take on Darwinism is so insightful that I shall quote it in full:

> It is always a warning indicator when a scientific theory plays a greater role outside its field of application than within it. Fascinating as Darwin's account of *The Origin of Species* (1859)

was, its real contribution lay beyond the explicit reference of the study. More important than understanding the stratification of the emergence of species, and even than the evolutionary mechanism propounded to explain the emergence, was the function of Darwin's theory in constituting a worldview. It was welcomed and repulsed for the exact same reason. Darwin had shown how creation could dispense with a creator. A world of chance developments could, over a sufficiently long period of time, evolve into a world of order. It was not the suggestion that men were descended from apes that was the most shattering realization, but that everything had originated through the survival of the best adapted random mutations. The most compelling natural indication of a supreme intelligence—the argument from design—had been decisively undermined. With such large theological reverberations, it was no wonder that Darwin's theory of biological evolution should receive scant attention on its own merits. It is a situation that prevails virtually up to the present.

Darwinian evolutionism functions to such an extent as the overarching worldview of modernity that even its subjection to scientific analysis is treated with deep misgivings. Everyone is more comfortable if its examination is reduced to the stylized opposition between evolution and creationism. That way, no one has to pay serious attention to the minor consideration that neither of them can be taken seriously as scientific theories. They cannot be disproved because the theories are designed to accommodate all contrary or missing evidence against them. This would be no more than a harmless intellectual idiosyncrasy if it did not have such disastrous consequences for science. Like counterfeit, the problem is that bad science drives out the good. Even today it is virtually impossible for conscientious biologists to admit that the evidence for evolution is extraordinarily thin. We simply have little tangible proof that one species evolves into another. As Darwin recognized, the fossil record, which is

> ultimately the only conclusive indication, is the weakest source of support. We have neither experience nor evidence of intermediate forms. It is clear that different species emerged and disappeared at different times, just as it is clear that chemical and genetic continuities are present across species. But the incubus of evolutionism hangs as such a dead weight on the scientific mind that even the best efforts to consider its revision encounter levels of resistance out of all proportion to their content. No one dares to attempt the removal of the ideological carcass from fear of the consequences of universal disapproval. More often than not, the voices of dissent come from outside the biological community. One wonders what force holds such regressive formalism in place. The only suggestion is that the anti-theological significance of evolutionism as a worldview continues to outweigh its scientific value. By calling into question the Darwinian universe, we would at the same time be restoring the openness to the transcendent creator. It is in other words the fear of God that prevents the biological community from too openly discarding a theory they have long ceased to honor in practice.

Commentary could only diminish the wisdom of Walsh's words, so I will turn to the second thing I want to do in this section.

In a book that cuts as wide a swath as mine does, it is important to have hotlines to specialists in the fields touched on, and my line to Darwinism connects with Jonathan Wells. Wells earned a Ph.D. in theology at Yale University, writing his doctoral dissertation on nineteenth-century Darwinian controversies. His research convinced him that the conflict between Christianity and Darwinism revolves around the issue of design. Christianity affirms that human beings were created in the image of God, while Darwinism claims that human beings were accidental by-products of an unguided natural process.

Not content with merely pinpointing the source of the conflict, Wells decided to pursue graduate study in biology. He earned a second Ph.D. at the University of California, Berkeley, specializing in embryology and evolution. Having examined the evidence on which Darwinism is based, he has become an outspoken critic of it. As a result, he has been coming under increasing attack by Darwinists. Wells, though, is used to controversy. In the 1960s, he spent a year and a half in prison for refusing to cooperate with the U.S. Army during the Vietnam War.

Concerned that standards in science and biology be maintained, Wells has written a book, *Icons of Evolution,* that exposes the fraudulence in continuing to include in biology textbooks illustrations that conflict with published evidence that biologists have known for years, with students given no indication that the icons are fraudulent.

One of those icons is the 1953 Miller-Urey experiment, which used a simulated primitive atmosphere to produce some of the molecular building-blocks of life. But geochemists have been convinced for decades that the earth's primitive atmosphere was nothing like the Miller-Urey simulation and that the experiment's findings have little or no bearing on the origin of life.

Another famous image is the Darwinian tree of life, according to which all modern species evolved gradually from a universal common ancestor. But the fossil record shows that the major groups of animals appeared together, fully formed, with no evidence of common ancestry—a pattern exactly opposite to Darwin's prediction.

Still another image is a set of drawings by Ernst Haeckel showing similarities in vertebrate embryos that supposedly point to common ancestry. But biologists have known for over a century that Haeckel faked the similarities and that early vertebrate embryos are quite different from each other.

These and other textbook misrepresentations cast serious doubt on what Darwinists claim as evidence for their theory. Wells acknowledges that Darwinian evolution works at some levels, such as antibiotic resistance in bacteria and minor changes in finch beaks. But he notes that evidence is lacking for the theory's larger claims. In particular, Wells insists that the Darwinian claim that humans are by-products of unguided natural processes is not a scientific inference, but a philosophical doctrine.

Cognitive Psychology

If physics is the fundamental and oldest science, cognitive psychology is the youngest. At first glance it looks like a throwback to crude materialism, for neuroscience (the cornerstone of cognitive psychology) is in its adolescence, and the field is drunk with its dizzying growth and the prospect of limitless horizons. (On the day that I wrote these words an alumni couple announced that they were giving MIT sixty-five million dollars for brain research!) This has brought the return of mental materialism. Not only is it back; it is back with a vengeance, in an unapologetic, out-of-the-closet, almost exhibitionistic form.

What makes cognitive psychology interesting is what is happening on another flank, the mind-body problem. Colin McGinn's book *The Mysterious Flame* presents most engagingly what I shall describe, so I will use him and his book as reference points.

The mind-body problem was foisted on the world by René Descartes, who split the world into mind and matter. He used God to bridge the two halves, but that resource is not available to scientists (or philosophers) anymore, and the residual bridgeless gap between mind and brain constitutes the mind-body problem.

The problem itself is easily described. We have minds (consciousness) and we have bodies (in this context, brains), neither of which can be converted into the other. Equally obvious is the two-way relationship that exists between them. If my mind orders my right forefinger to type the letter *J*, it obeys; from the other side, if my brain grows tired from obeying such orders for several hours, I *feel* tired. The problem is, how can neuron firings in my brain give rise to things as different from them as are my thoughts and feelings? And vice versa.

The scientists and philosophers I am considering—to McGinn's name I should add those of Thomas Nagel at New York University and Stephen Pinker, who heads the cognitive science program at MIT—give their position on the mind-body problem the awkward name *mysterianism.* That label reflects the frank admission that in the three centuries since Descartes set the problem in place, not one iota of progress has been made toward resolving it. McGinn (who as Pinker rightly says "thinks like a laser and writes like a dream") dramatizes the impasse by quoting a clever excerpt from a science fiction story by Terry Bisson. An alien explorer, just returned from an earth visit, is reporting to his commander:

> "They're made out of meat."
>
> "Meat?" . . .
>
> "There's no doubt about it. We picked several from different parts of the planet, took them aboard our recon vessels, and probed them all the way through. They're completely meat."
>
> "That's impossible. What about the radio signals? The messages to the stars?"
>
> "They use the radio waves to talk, but the signals don't come from them. The signals come from machines."
>
> "So who made the machines? That's what we want to contact."
>
> "They made the machines. That's what I'm trying to tell you. Meat made the machines."

"That's ridiculous. How can meat make a machine? You're asking me to believe in sentient meat."

"I'm not asking you, I'm telling you. These creatures are the only sentient race in the sector and they're made out of meat."

"Maybe they're like the Orfolei. You know, a carbon-based intelligence that goes through a meat stage."

"Nope. They're born meat and they die meat. We studied them for several of their lifespans, which didn't take too long. Do you have any idea of the life span of meat?"

"Spare me. Okay, maybe they're only part meat. You know, like the Weddilei. A meat head with an electron plasma brain inside."

"Nope, we thought of that, since they do have meat heads like the Weddilei. But I told you, we probed them. They're meat all the way through."

"No brain?"

"Oh, there's a brain all right. It's just that the brain is made out of meat!"

"So . . . what does the thinking?"

"You're not understanding, are you? The brain does the thinking. The meat."

"Thinking meat! You're asking me to believe in thinking meat!"

"Yes, thinking meat! Conscious meat! Loving meat. Dreaming meat! The meat is the whole deal!"

The mysterians, having gotten us where they want us—which is to see that science has made no progress at all in allaying the absurdity of the concept that squishy gray matter in our heads ("thinking meat") can cause mental life whereas similar-appearing liver meat cannot—they then unload on us their surprise: we may be stuck with this problem for as long as we human beings are around to ponder it. For what do we think we are, they ask—

omniscient? Every day we discover anew that the world is more strange, more complicated, and more mysterious than we had suspected. This leads mysterians to speculate that the mind-body problem may be just too big for us to get our finite minds around.

This is a new note to hear from science. It gives us not just a novel answer to a problem, but a novel *kind* of answer—one that is refreshingly different from the standard "Give us time and money and we'll deliver the goods." I must not press the difference too far. McGinn and company are not throwing up their hands in despair. What they are intent on doing is uncovering the deep reasons for our bafflement regarding the problem at stake.

When I first encountered McGinn's line of thought, I realized that I had heard something like it before, and I quickly traced it to a lecture I had heard Noam Chomsky deliver during one of my last years at MIT. (McGinn acknowledges his debt to Chomsky and others.) No other living species has the endowment for language that human beings have, Chomsky said, but every species seems to have comparably distinctive ways of understanding the world and managing to relate to it. Birds are born with an instinctive talent for knowing how to build nests, one that we could not match if we gave our entire lives to the project. Ants have a knack for working together instinctively in building anthills; each ant coordinates its activities with those of other ants in its colony in ways that put human task forces to shame. As for us humans (Chomsky rounded out his lecture saying), we are obviously good at language and science. No other species can rival us on those fronts. The downside of all this is that while being good at some things, every species is poor at others. Years later, the mysterians are continuing this point.

The question here is whether the mysterian thesis bears on the issue of light at the end of the tunnel. Toward the close of *The Mysterious Flame,* McGinn says that his book cannot escape the conclusion that "a radical conceptual innovation is a prerequisite for solving the mind-body problem. . . . It requires two new concepts, one for the mind and one for the brain." To which I will myself add that (given the way that parts reflect the wholes they are parts of) those two new concepts require a new worldview—which is to say, one that differs from the scientific one we now have.

Thomas Nagel himself said as much in his contribution to the conference that the CIBA Foundation convened in London in 1992.

> The apparent impossibility of discovering a transparent connection between the physical and the mental should give us hope, for apparent impossibilities are a wonderful stimulus to the theoretical imagination. I think it is inevitable that the [search for the link between mind and brain] will lead to an alteration of our conception of the physical world. In the long run, therefore, physiological psychology should expect cosmological results. Physical science has not, heretofore, tried to take on consciousness. Now that it is doing so, the effort will transform science radically.

My inclination is to turn this around. How would things look if we said, Consciousness has not, heretofore, tried to take on physical science? Yet this is the one logical possibility that the mysterians never consider—or rather, never consider *seriously,* for McGinn does touch on it, but only to dismiss it.

We shall hear more of this possibility in my final chapter. Here I want only to raise a second possibility that seems not to have occurred to the mysterians: If the human mind is mysteriously endowed with an innate talent for science (the mysterians' field), what rules out the possibility that it is equally mysteriously endowed with a talent for knowing the Big Picture (my field)? That picture is the topic of Chapter Fourteen.

CHAPTER 12

TERMS FOR THE DÉTENTE

Thinking back over the past two decades, it occurs to me that I have had the extraordinary opportunity of sharing the platform with four world-class scientists: David Bohm in quantum mechanics, Carl Sagan in cosmology, Ilya Prigogene in chemistry, and Karl Pribram in neuroscience. None of the four (not even David Bohm) would grant me that the scientific method is limited.

This is the kind of misreading of science that got us into the tunnel in the first place, for it belittles art, religion, love, and the bulk of the life we directly live by denying that those elements yield insights that are needed to complement what science tells us. This is like saying that the important thing about a human being is her skeleton as it shows up on X-ray plates. Our exiting the tunnel requires that science share the knowledge project equitably with other ways of knowing—notably (in this book) the ways of God-seekers.

What *equitably* means here is the question for this chapter, and I begin with the premise that science must move over. Eventually the sun sets on every empire, and the sunset for the empire of science has arrived. That it should continue as an honored partner in the knowledge quest goes without saying, and (because this is turning out to be a more personal book than I had anticipated), I shall be frank about my personal reasons for honoring science. Seven years ago my physician discovered that my PSA count was off the charts, a sure sign of prostate cancer. (When I reported this to my family, adding that I had thought that PSA stood for Pacific Southwest Airlines, now defunct, a daughter who was visiting told me, not for the first time, "Father, you are more innocent than anyone your age is entitled to be." But on with my story.) A urologist teamed up with an oncologist, and already their joint ministrations have given me five more years of life than I would otherwise have enjoyed. If I did not sincerely thank and honor science for this gift, I would be a colossal ingrate.

That said, to understand how far (and in what ways) science must move over requires that we understand what science is. I defined the word provisionally in the chapter on scientism, but the time has come to be more precise.

Semanticists tell us that we can use words in any way we please as long as we are clear about our definitions and stick to them. Nowhere in this scientistic age is that liberty more exploited than in the case of "science." Truly bizarre claims are made in the name of this god, but those claims make perfect (logical) sense if science is what their authors say it is. David Bohm, whom I just mentioned, provides me with a good example of this. When he surprised me by denying that science is limited and I asked how he defined the word *science,* he answered, "Openness to evidence." When I said that definition made me a scientist, he came back, "Perhaps you are." Language breaks down with such an answer.

I do not want to leave David Bohm saddled with what I just said, for he is one of my heroes and was a personal friend as well, for I brought him to Syracuse University for three weeks while I was there and (during those weeks) served as his host. If it were only to explain how his inclusion of me as a scientist is less bizarre than it sounds, I would not devote the following section to the man; but in accomplishing that purpose, the story I am about to tell will advance in additional ways this chapter's search for what science should be taken to be.

A Glimpse of David Bohm

At one point during my decade at Syracuse University the administration entered a line-item in its budget to enable the humanities division to bring to the campus each year for three weeks a distinguished visiting professor of humanities. I was appointed to chair the search committee, which consisted of one member from each of the division's five departments.

Saul Bellow for the English Department was an easy choice, as was Noam Chomsky for the philosophers. Next it was the religion department's turn, and (as its representative) I put forward the name of David Bohm. Pandemonium! "You know that the administration gave us this sop to salve its conscience for shortchanging the humanities, and you propose that we give the plum to a *scientist!*" they protested. When the hubbub died down to the point where I could be heard, I admitted that I was indeed doing that, but that I had my reasons. Bohm's doctrine of the implicate order that transcends space and time housed more important implications for religion than anything any religious studies professor we could think of was saying. The committee was not mollified, but I had voted for their candidates, so they had no choice but to vote for mine.

Academic protocol requires that if you officially invite to your campus someone in another field, you clear the invitation with the department in question, so before we invited Bohm I went to the chairman of our physics department to secure his approval. He was ecstatic at the prospect. "Everyone in our department cut his quantum mechanics teeth on Bohm's textbook," he said, and they would all be overjoyed to have him on campus. Could they have him for one of their departmental colloquia? As I was leaving his office, the chairman followed me into the hall to assure me that if I needed more money, just to let him know.

Bohm accepted our invitation, and in due time he arrived for his visit. His three-week stay opened with a Monday evening lecture for the general public. The physics department was out in force.

The physics colloquium took place two days later. When Bohm and I arrived at the departmental office, the chairman welcomed him and then turned him over to several senior professors in order to draw me into the hall. "Huston," he said, "I want to let you know that he will not have a friendly audience." Things Bohm had said in his Monday address had not sat well with the physicists.

When it was time to proceed to the colloquium, we found our way blocked by mobs of faculty and students in the corridors. A backup was in place, and word was circulated that we would proceed to room such-and-such. It too proved inadequate, and what was to have been a colloquium ended as a lecture in the largest hall in the physics building. Even so, some students had to stand throughout the event.

Once introduced, David Bohm mounted the large stage and (without glancing at a note the entire time) talked nonstop for an hour and a quarter as he paced back and forth, covering the three-section, three-tier blackboard with incomprehensible equations. Glancing around the hall, I suspected that within ten minutes he had lost everyone but a handful of senior professors, but he kept on talking. And the audience kept on listening, if for no other reason than to remember for the rest of their lives the experience of watching the workings of the mind of the man who had worked closely with Einstein and whose Hidden-Variable Theory continued to hold out a (minority) hope that Einstein was right in thinking that God does not play dice.

When finally, as abruptly as he had begun, Bohm stopped talking and sat down, the chairman called for questions. Instantly the arm of a senior professor in the front row shot up. "Professor Bohm," the questioner said, "this is all very interesting philosophy. But what does it have to do with physics?" I glanced at the solid bank of equations that stared out at us from the blackboards, with not a single *word* in sight. Without batting an eye, Bohm replied, "I do not make that distinction."

A pall fell over the hall. One or two polite questions brought the afternoon to a close.

I said that I was including this recollection on David Bohm in part to redeem his seeming naivete in defining science so broadly as to include me, and his retort to his questioner does that; for if you do not separate science from philosophy, it does indeed follow that science is unbounded. Whether ultimately the two *can* be separated is too large a question to be entered into here, so I go back now to pick up the question I had begun before David Bohm diverted me. How can science best be defined for purposes of public discourse today?

Science Rightly Defined

In the chapter on scientism I offered my definition of science, which (forgoing frills) is this: Science is what replaced traditional societies with the modern, technological, industrial world. What accomplished that transition was the controlled experiment. Science is the body of facts about the natural world that controlled experiments require us to believe, together with logical extrapolations from those facts, and the added things that scientific instruments enable us to see with our own eyes.

Regularly I find this definition faulted for defining science in its narrowest, most hard-nosed sense—which is precisely my intent, for any looser definition points toward the tunnel. Only thus narrowly defined does science tell us what we *must* believe. Every enlargement of the definition produces cracks into which philosophy can seep to weaken the claims put forward. Because philosophy always allows for reasonable differences, the claim that a scientific hypothesis makes on us weakens in direct proportion to the increase of philosophy in the mix.

This leaves us with a choice. Either we restrict the word *science* to what *must* be believed (which requires my narrow definition), or we relax the way we define it and demote its truth-claims to suggestions backed by sliding scales of reasonableness. Because the second option—science as suggestions—contravenes our public understanding of the enterprise (which understanding is on the right track, for suggestions could not have created our technological, industrial world), it fosters the confusion that besets us. Mine is the option that renders clear thinking about science possible.

That paragraph is so basic to the arguments of this book that I encourage the reader go back and reread it.

The Limits of Science

The television program that the British Broadcasting Company mounted for the centennial of Einstein's birth was brilliant throughout, but nothing matched its two opening sentences: "Einstein would have wanted us to say it in the simplest possible way. Space tells matter how to move; matter tells space how to warp." That is more than just brilliant. In the way it cuts through technicalities to get to the point, it is helpfully brilliant. Schooling myself to that opening, I offer this as my counterpart for the book in hand:

Two worldviews, the traditional and the scientific, compete for the mind of the third millennium. (E. O. Wilson's wording of this first of my two sentences is, "The choice between transcendentalism and empiricism will be the coming century's version of the struggle for men's minds.") If we had our choice, we would prefer the traditional worldview; and we do have that choice, because neither of them can be proved to be truer than the other.

The support for that last assertion lies in understanding science's limitations, for only if we have those clearly in mind can we see that science has no lien on the traditional outlook. Science obviously has a better grasp of the calculable features of the physical universe, but whether those features comprise all that exists cannot be scientifically determined.

These bare bones of the matter were laid out in Chapter Three, but because our culture nods (in the dual sense of "agreeing to them" and then "dozing off" and forgetting what it just assented to), I propose here to bear down hard on this point by way of an image, an anecdote, and then a full-scale analysis.

The image. Imagine yourself in a bungalow in North India. You are standing before a picture window that commands a breathtaking view of the Himalayan Mountains. What modernity has done, in effect, is to lower the shade of that window to within two inches of its sill. With our eyes angled downward, all that we can now see of the outdoors is the ground on which the bungalow stands. In this analogy, the ground represents the material world—and to give credit where credit is richly due, science has shown that world to be awesome beyond belief. Still, it is not Mount Everest.

The anecdote. In his *Guide for the Perplexed,* E. F. Schumacher tells of getting lost while sightseeing in Moscow during the

Stalinist era. As he was puzzling over his map, an Intourist guide approached him and pointed on his map to where they were standing. "But these large churches around us," Schumacher protested; "they're not on the map." "We don't show churches on our maps," the guide responded crustily. "But that can't be," Schumacher persisted. "The church on *that* corner is on the map." "Oh, that," said the guide. "That *used* to be a church. Now it's a museum."

Precisely, Schumacher goes on to say. Most of the things that most of mankind has most believed in did not show on the map of reality that his Oxford education gave him. Or if they did, they appeared as museum pieces—things that people believed in during the childhood of the human race but believe in no more.

That anecdote and the image that preceded it are intended to drive home the fact that (in the process of showering us with material benefits and awesome knowledge of the physical universe) science has erased transcendence from our reality map. (Remember the blunt statement quoted in Chapter Two from the *Chronicle of Higher Education:* "If anything characterizes 'modernity,' it is the loss of faith in transcendence, in a reality that encompasses but surpasses our quotidian affairs.") I proceed now to how such erasure takes place. This requires moving from allusions to arguments.

The Analysis. I shall present my argument regarding science's limitations here in two forms. Each contains six numbered propositions that give the appearance of elongated syllogisms. The first form runs like this:

1. Science (as I have already quoted Alex Comfort as saying) is our "sacral" mode of knowing. As the court of final appeal for what is true, it occupies today, quite isomorphically, the place revelation enjoyed in the Middle Ages. An intellectual historian has written that already a hundred years ago, Westerners had come to believe more in the periodic table of chemical elements than they believed

in any of the distinctive things the Bible speaks of—angels, miracles, and the like.

2. The crux of modern science is the controlled experiment. This explains our confidence in science (as noted in Point 1), for such experiments winnow true from false hypotheses and thereby offer *proof.*

Now watch this next point, for in the context of the two that precede it, it is one of the few original ideas in this book. (That, at least, is how it looks to me. Most of the book deals with things we already know yet never learn, but when this thought popped into my mind, it brought the *Eureka!* sensation that attends original insights.)

3. We can control only what is inferior to us. To make this point clear, I need to sharpen my terms a bit. I mean *intentionally* control, for if I locked myself out of my house, its walls would thwart my wish to enter without their being my superior. And I am speaking of control *across special lines,* for within the same species variables can produce exceptions. (The Nazis controlled the Jews but were not superior to them.) By *superior* and *inferior* I mean to invoke every criterion of worth we know, and perhaps some we do not know. Galaxies are bigger than we are and earthquakes pack more power, but we know of nothing (empirically speaking) that is more intelligent and free than we are or more compassionate than we can be. It seems apparent that human beings have controlled the American buffalo more than vice versa. It is that kind of correlation between control and orders of existence that this third point calls attention to.

With these three points in place, the fourth follows automatically.

4. Science can register only what is inferior to us. Ask yourself if, in any science course you have taken or any science textbook you have been assigned, you have ever been informed of something that exceeds us in our distinctively human attributes. Those who resist the

negative answer to this question will try to convert its rhetorical "no" into "not yet," but the force of the preceding point is to show logically—which is to say, *in principle*—why the substitution will not work. Try to imagine for a moment what beings superior to ourselves might be. Disembodied souls? Angels? God? If such beings exist, science—the science that can prove its propositions through controlled experiments—will never bring them to view for the sufficient reason that if they exist, it is they who dance circles around us, not we them. Knowing more than we do, they will walk into our experiments if they choose to; otherwise not. Karl Pribram (who did as much to popularize the hologram as anyone) tells me that it takes about seven years now to mount an important experiment on the brain; it takes that long to determine what the relevant variables are. In the case of beings more intelligent than ourselves, we have no clue as to how their minds work, so there is no way to discover what variables would be needed to mount an experiment on them.

Because the resistance to this point is in proportion to its importance—the point being, to repeat, that science can disclose only what is inferior to us—I shall dwell on it for another couple of paragraphs.

If we liken the scientific method to a flashlight, when we point it down toward the path we are walking on its beam is clear and strong. Suppose, now, that we hear footsteps. We want to know who is approaching and raise our flashlight to horizontal position. (This represents turning the scientific method on our equals, our fellow human beings, and shifting from the natural to the social sciences.) What happens? The flashlight develops a loose connection. Its light flickers and we cannot get a clear image. Freedom makes human beings difficult to corner experimentally. No one (a social scientist has written) knows why crime occurs, why marriages break down, why war occurs, why economies fall into depression, or why governments cannot eradicate corruption.

As for psychology, it can tell us a few things about people in the aggregate, but individuals in their existential uniqueness (to say nothing of their souls and spirits, if there be such) it cannot get to. To lock this into the present point (which again is that science can disclose only what is inferior to us), it is axiomatic in the social sciences that in experiments that involve human subjects, the subjects must be kept in the dark about the experimental design, which of course places them in an inferior position respecting the experimenter, who knows what is going on.

Finally (to complete this analogy), if we tilt our flashlight skyward—or toward the heavens, as we might appropriately say here—the batteries in our defective instrument clunk to the bottom of the casing and the light blacks out. This, of course, does not prove that there *are* things in the sky. But it does argue that *if* there are, science will not discover them.

The two final entries in this first argument belong together, for the first of them merely recapitulates the heart of what has preceded to prepare the stage for the argument's conclusion.

5. Because we take our clues from science as to what exists (Point 1), and science can disclose only what is inferior to us (Point 3), it follows that:

6. We are trying to live superior lives (the best we can make them) in an inferior world. Or, if you prefer, *complete* lives in an *incomplete* world.

That first argument brought to light one limitation of science and the consequence if we give in to it. This second one enters the full list.

There are six things science cannot get its hands on:

1. *Values in their final and proper sense.* Close friends at the start, Bertrand Russell and Ludwig Wittgenstein ended at opposite ends

of the philosophical spectrum, but on one point they remained in full agreement: science cannot deal with values. Russell proposed one exception—except insofar as science consists in the pursuit of knowledge—but that is not really an exception, for although that value is assumed by scientists, it is not itself scientifically derived. Science can deal with *instrumental* values, but not *intrinsic* ones. *If* health is valued over immediate somatic gratification, smoking is bad, but the intrinsic values that conflict (health versus pleasure) science cannot weigh. Again, science can deal with *descriptive* values (what people *do* like) but not *normative* ones (what they *should* like). Market research and opinion polls are sciences; indeed, when the margins for error are factored in, they come close to being exact sciences. As such, they can tell us whether people prefer Cheerios to Raisin Bran and who is likely to win an election. Who *should* win is a different story. There will never be a science of the *summum bonum,* the supreme good.

2. *Existential and global meanings.* Science itself is meaningful throughout, but on existential and global meanings it is silent. *Existential* meanings are ones that concern us; they relate to what we find meaning-full. Scientists can spread before us their richest wares; but if the viewer is depressed and buries his head in his arms, scientists cannot compel his interest. (Prozac only muddies the water here, so I will let it lie.) *Global* meanings are of the sort, *What is the meaning of life?* or *What is the meaning of it all?* As human beings, scientists can invest themselves in these questions, but their science will not help them find answers to them.

3. *Final causes.* For science to get on with its job, Aristotle's final causes—the *why* of things—had to be banished and the field left to explanations by way of efficient causes only. Except in biology, we must add. Living creatures seek food and sex to satisfy their hunger and libidinal drives, and their satisfactions are the final cause of their hunting. (Tolman's *Purposive Behavior in Animals and Men* was a highly regarded book in my college days.) So *teleonomy,* yes,

but *teleology* (final causes outside the animate world), no. Whether the case be that of Galileo's falling rocks or Kepler's light, the shift from classical to modern mechanics was brought about by the separation of primary from secondary qualities—which is to say, the separation of nature's *quantitative* from its *qualitatively experienced* features. Talk of volition and the why of things was removed to let impersonal laws of motion take over. Near the start of modern science, Francis Bacon stated this with characteristic vividness. He likened teleological explanations in science to virgins dedicated to God: "barren of empirical fruit for the good of man."

4. *Invisibles.* Here too a qualification must be inserted. Science can deal with invisibles that can be logically inferred from observable effects. In the early 1900s, Michael Faraday discovered magnetic fields in this way by placing iron filings on a piece of paper and a magnet underneath. When he vibrated the paper slightly, lines of magnetic force appeared. The randomly scattered filings fell into lines as if ordered by a drill sergeant, revealing the pattern of the magnetic field. But if there are invisibles that do not impact matter thus demonstrably, science gets no wind of them.

5. *Quality.* Unlike the preceding four, this fifth exclusion does not need to be qualified. And it is basic to the lot, for it is the qualitative ingredient in values, meanings, purposes, and noninferable invisibles that gives them their power. Certain qualities (such as colors) are connected to quantitative substrates (light waves of given lengths), but the quality itself is not measurable.

6. *Our superiors.* This was covered in the initial six-point argument.

Division of Labor

When we put together the six things science cannot deal with—simplified to help us keep them in mind, they are values, meanings, final causes, invisibles, qualities, and our superiors—we see that science leaves much of the world untouched. A division of labor suggests itself. Science deals with the natural world and religion with the whole of things, as this diagram suggests:

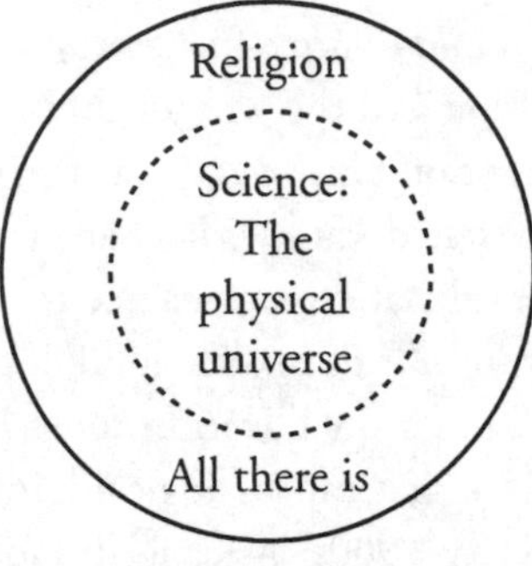

Figure 1.

That religion is represented by the larger of the two circles seems to give it the advantage, but that impression is corrected when we note that science works more effectively with its part than does religion with its. Science houses precise calculations, knockdown proofs, and technological wonders, whereas religion speaks in generalities, such as "In the beginning God created the heavens and the earth," or "The heavens declare the glory of God," or "All things are the Buddha-nature," or "The world is maya," or "Only heaven is great." Oliver Wendell Holmes's way of establishing parity is appealing: "Science gives us major answers to minor questions, while religion gives us minor answers to major questions."

If this way of slicing the pie is accepted, it follows that both parties should respect the other's sphere of competence. It would be unrealistic not to expect border disputes to erupt; but they should be negotiated in good faith without losing sight of the terms of agreement. When scientists who are convinced materialists deny the existence of things other than those they can train their instruments on, they should make it clear that they are expressing their personal opinions like everybody else and not claim the authority of science for what they say. From the other side, religionists should keep their hands off science as long as it is genuine science and not larded with philosophical opinions to which everyone has rights. All responsible citizens have a right to oppose harmful outcomes that some scientific research could lead to—germ warfare, cloning, and the like—but that is an ethical matter, not one that relates to science proper.

I am not so naive as to think that even if my proposal were accepted it would make for a just and durable peace. I do believe, however, that it points in the right direction. What is most right about it is that it allots religion an ontological domain of its own. It proposes respect for religion's concern to posit and work with things that exist objectively in the world but which science cannot detect. I find this underplayed in the current dialogue, where theologians too often accept science's inventory of the world as exhaustive and content themselves with discerning the meaning and significance of what science reports.

The Cow That Stands on Three Legs

To conclude this chapter I will follow Peter Drucker in an earlier chapter and indulge myself in a wishful fantasy.

In the California Bay Area, I sit alongside a tripod of three institutions dedicated to the science-religion issue. At the Graduate Theological Union in Berkeley there is Robert Russell's Center for Theology and the Natural Sciences; at the California Institute for Integral Studies in San Francisco there is Brian Swimme's Center for the Story of the Universe; and in Sausalito there is the (recently deceased) Willis Harmon's Institute for Noetic Sciences. All were founded by scientists, and (as I often say to myself petulantly) they are all doing it wrong. By which, of course, I mean that they are not handling the science-religion issue in exactly the way I think it should be handled—a classic case of Freud's "narcissism of small differences." (He did write well.)

The situation makes me think of Hinduism's theory of the four *yugas*—four steadily declining ages that occur in every cosmic cycle. India likens the decline to a cow, which in the first age stands solidly on all four legs, in the second limps on three, in the third wobbles on two, and in the last teeters on a single leg before collapsing, whereupon the cycle starts all over again. Living within a stone's throw of the three above institutions makes me feel as if I am living in the three-legged *treta yuga*.

The Center for the Story of the Universe wants to awaken people to how mind-bogglingly stupendous, glorious, awesome, and precious the universe is. The Institute of Noetic Sciences wants to enlarge science by turning its spotlight on issues (such as alternative medicine) that have not been sufficiently investigated. The Center for Theology and the Natural Sciences wants to dialogue with scientists to discover ways in which each can learn from the other.

All three are important projects, so what could be so bad? Why am I so irritable? Because three well-functioning legs do not make for a robust cow.

If I am right in thinking that *the* greatest problem the human spirit faces in our time is having to live in the procrustean, scientistic worldview that dominates our culture, there is an urgent need for a fourth center that would dedicate itself to extricating the spirit from that cage. Named, perhaps, the Equal Opportunity Center for Science and Religion (EOCSR), it would have two main departments.

The first would act as a watchdog on scientism. Keeping its eye peeled for places where science modulates into scientism, this department would flag those unwarranted movements in a monthly, four-page publication that would point them out, with equal space provided for rejoinders from the "accused." When, for example, Richard Dawkins and Steven Pinker teamed up recently to mount a discussion on "Has Science Killed Spirit?" and concluded that the answer is no if *spirit* means zest and yes if it refers to a homunculus in the head, this department's watchdog newssheet would flag the cheap shot in the allusion to homunculae and go on to note that empirical studies are methodologically incapable of determining whether extra-epiphenomenal invisibles do or do not figure in the workings of the brain. And when, during the question period of that hugely attended London/Paris event, Dawkins answered the charge that his position was reductionistic by saying, "Reductionism makes me want to reach for my revolver, for there is no such thing," the newssheet would point out that that is like saying, "I'm not reducing the upper stories of a skyscraper to its ground floor because there are no upper stories."

The other arm of the Equal Opportunity Center for Science and Religion would mount an ongoing series of monthly discussions on issues where scientific and religious understandings appear to conflict, the obvious ones at present being Darwinism and intelligent design. These are already being vigorously discussed, so what the EOCSR would offer here is a different format. Technical papers (if speakers brought them) would be available on a table for anyone who wished to take one home, but the programs would take the form of discussions chaired by a hardboiled, fair-minded judge who would hold the speakers (not more than two in any given evening) to alternating ten-minute statements until, after forty minutes, the floor would be opened to the audience. The speakers would be encouraged to question their opponents as much as state their own positions. If the discussion wandered, the chair would bring it back by asking, "What is the issue?" The full spectrum of positions on the issue in question would be allowed a hearing, even short-term creationism on the evolution front. The emphasis throughout would be on effective communication aimed toward educating the interested public. Fancy footwork to prove how knowledgeable the speaker is would be scathingly dealt with.

It would be good if the EOCSR could be housed in a theological seminary, for nothing is more important for the future of the church than that its servants be solidly grounded in the issue—science versus religion—that holds the fate of the church in its hand.

CHAPTER 13

THIS AMBIGUOUS WORLD

Perception, as we now know, is a two-way process. The world comes to us, and we go to it—with inbuilt sensors, concepts, beliefs, and desires that filter its incoming signals in ways that differ in every species, every social class, and every individual. In a way, we share the same world with birds, and speak blithely of a bird's-eye view of it, but what such a view would look like we have no idea.

What concerns us here is the way our concepts, beliefs, and desires affect worldviews. As the title of this chapter announces, the world is ambiguous. It does not come tagged "This is my Father's world" or "Life is a tale told by an idiot." It comes to us as a giant Rorschach inkblot. Psychologists use such blots to fish in the subterranean waters of their patients' minds. The meanings latent in the shapes of the blots on the ten cards are not inscribed on them. The blots approach the patient as invitations: *Come. What do you see here? What do you make of these contours?*

LIFE'S COSMIC INKBLOT

The sweep of philosophy, written and unwritten, supports this inkblot theory of the world conclusively. People have never agreed on the world's meaning, and (it seems safe to say) never will. Anthropologists tell us that even in tribes so small and isolated that one might expect a unanimous outlook, the village atheist still turns up. He may keep his dissident views to himself and sit out the rituals or move through them perfunctorily. But he is there, reading the cosmic inkblot in his own heretical fashion. We talk of the mystical East and spiritual India, but India has an atheistic, materialistic, hedonistic, *charvaka* tradition that goes way back. Its watchword, "Life is short, so I shall eat butter," is India's counterpart to "Eat, drink, and be merry, for tomorrow we die."

There is something in the human makeup that resents this state of things. Why are we forced to grope our way to life's meaning and discover it for ourselves? Why can't we just be *told* the fact of the matter? Kierkegaard is useful here. He tells us that although we *think* we would like to be told, if we *were* told we would not like the position that would place us in. For it would deprive us of our freedom and thereby our dignity, leaving us robots. All that would remain for us to do would be to look up the answers to our questions in life's answer-book and apply them mechanically to our problems.

Our actual condition is the opposite of this. Instead of flunkies, we are free agents. In the most arresting phrase in his writings, Kierkegaard tells us that we have been given the freedom to "choose ourselves." Buddhists are adamant in insisting that of the six kinds of beings (gods, jealous gods, hungry ghosts, hell-beings, animals, and humans), humans are the most fortunate of the lot, for only they possess the one thing that can release creatures from the relative, *samsaric* world—namely, free will. An answer-book would deprive us of the greatest power we have in life—the power to decide what we want to do with our lives, what we want to give them to. That decision is sometimes made in a single swoop and sometimes forged incrementally (and almost unnoticed) by the daily micro-decisions life requires of us. For life comes at us fired point-blank, as Ortega y Gasset put the matter. It does not ask, Are you ready to get married? Do you know enough to have children? Here it comes, ready or not, requiring that we decide.

Wow he died as wow he lived,
Going whop to the office and blooie home to sleep and
biff got married and bam had children and oof got fired,
zowie did he live and zowie did he die.

These lines from Kenneth Fearing articulate how life feels to us most days. Only as we look back at the course our footprints have traced do we fully understand how we have read life's inkblot.

Along with multiculturalism, which has faiths rubbing shoulders as never before, this recognition of the world's ambiguity could help to reduce the friction that has so bedeviled religious relationships in the past. (Cardinal Newman's anguished lament, "O how we hate one another for the love of God," echoes repeatedly in our ears.) My China background provides me with a window onto this possibility.

Quantitatively speaking, the Chinese Empire is the most impressive social organization human beings have ever created. When we multiply its duration (more than two thousand years) by the number of people this most populous nation on earth gathered under a single umbrella in an average year, it makes the empires of Alexander, Caesar, and Napoleon look episodic. (The Buddhist *sangha*, or monastic order, boasts a longer lifespan—twenty-five centuries compared with the twenty of the (now defunct) Chinese Empire—but its population is minuscule by comparison.) Part of the reason for China's success may lie in the way she positioned her religions as partners rather than antagonists. In the China I knew, if you asked people what church they belonged to, the typical answer would be, "The great church *(tai chao),* of course"—a federation of Confucianism, Taoism, and Buddhism woven together like strands in a single rope. As the then-going adage had it, every black-haired child of Han wears a Confucian hat, a Taoist robe, and Buddhist sandals.

That was my childhood. In my early years of teaching world religions, a student one morning turned up with a "Dear Abby" column from the *St. Louis Post-Dispatch.* The column had appeared on the day of our previous class session, in which I had presented China's distinctive way of positioning her religions. The column illustrated my point so vividly that I reused it every time the subject surfaced in my courses.

> Dear Abby,
>
> I am young, attractive, interested in religion and would like to get married. I belong to First Presbyterian Church, Blessed Angels Catholic Church, B'nai Amona Synagogue, and I attend Christian Science lectures regularly, though I do take aspirin occasionally.
>
> Can you tell me how I can meet a man who is interested in any or all of these religions?
>
> Ida

Abby's reply:

> Dear Ida,
>
> You seem to have the bases covered. I do not see how you can belong to all of those churches. . . .

Abby's letter continues for a few more sentences, but my point has been made. Naturally Abby could not understand Ida's multiple affiliations, for she was a Westerner. A Chinese would have had no difficulty.

I include this exchange not to suggest that the third millennium will (or should) become religiously syncretistic. An entire civilization went into making the East Asian formula work, and the West's "Choose ye this day whom ye shall serve" has its own merits which the next section of this chapter will touch on. But if we derive from the East Asian example not multiple affiliation but mutual respect, it seems quite possible that the new millennium could move in that direction.

A Sidewise Glance at the Social Scene

I am trying to keep this book focused on the Big Picture, for if I wander too far from that aim the book could easily degenerate into a gaggle of opinions on all manner of things. However, the human spirit is, correlatively, also the book's subject, and already at several points I have run into social developments that affect spirit so pronouncedly that it would seem contrived to skirt them. The glaring example at this point is the way that liberalism and conservatism have polarized religious America. The Islamic world is polarized too, but along different lines which I will not go into.

Generally speaking, religious conservatives regard the Truth by which they live as absolute and therefore appropriately capitalized, whereas liberals are more sensitive to its relativities—to the ways different points of view splinter the single, all-encompassing Truth and leave us with myriad lower-case truths. Both positions have their virtues and their limitations.

The downside of Truth is the danger of fanaticism. Because absolutes brook no alternatives, conservatives are tempted to invade their neighbors' autonomy and try to force Truth down their throats. Liberals face the opposite problem, for the danger that stalks relativism is that it will bottom out into nihilism. At that extreme, relativism collapses into the view that nothing is better than anything else. This is an unlivable philosophy, but the indiscriminate championing of tolerance has moved our society in its direction while debasing the meaning of tolerance in the act. The following passage puts this point more vividly than I could, so (with thanks to Michael Novak, my former colleague at Syracuse University, who crafted it) I will quote it:

> Tolerance used to mean that people of strong convictions would willingly bear the burden of putting up peacefully with people they regarded as plainly in error. Now it means that people of weak convictions facilely agree that others are also right, and anyway the truth of things doesn't make much difference as long as everyone is "nice." I don't know if "judgmentaphobic" is a word, but it ought to be. This republic crawls with judgmentaphobes. Where conscience used to raise an eyebrow at our slips and falls, sunny non-judgmentalism winks and slaps us on the back.
>
> In the absence of judgment, however, freedom cannot thrive. If nothing matters, freedom is pointless. If one choice is as good as another, choice is merely preference. A glandular reflex would do as well. Without standards, no one is free, but only a slave of impulses coming from who knows where.

That said, I turn to the happier side of the picture. Both liberals and conservatives also have their virtues. The virtue of liberalism is tolerance (in the valid, former sense of the word just indicated), and the virtue of conservatism (when likewise well handled) is the energy it can infuse into life through the feeling of certainty that the universe is on one's side.

This feeling can get drunkards out of ditches. One of the most arresting sentences I have come across in recent years brought with it a special moment in time—it drew me up short and caused me to put down the journal for several minutes to stop and think. The sentence read, "Liberals do not recognize the spiritual wholeness that can come from the sense of certainty." Embedded in that single sentence may be the primary reason that mainline liberal churches are losing ground to conservative churches. Liberals are at their worst in not recognizing how much an absolute can contribute to life, and in assuming that absolutes can be held only dogmatically, which is not the case. Absolutism and dogmatism lie on different axes. The first relates to belief, whereas the second is a character disorder. The opposite of absolutism is not open-mindedness but relativism, and the opposite of dogmatism is not relativism but open-mindedness. There can be, and are, dogmatic relativists and open-minded absolutists.

The preceding paragraphs show liberals as better than conservatives at recognizing the dangers of fanaticism and the virtues of tolerance, and conservatives as better at perceiving the dangers of nihilism and the virtues of the sense of certainty. Only one more step needs to be added, and it is an important one.

Both the strengths and the dangers of liberalism pertain to life's horizontal dimension, which encompasses human relationships (i.e., relationships between equals), whereas those of conservatives pertain to the vertical, asymmetrical God-person relationship. The sobering fact for religious liberals—the one that is causing them to lose ground to conservatives—is that, of the two dimensions, the vertical relation is the more important. It argues nothing against justice and compassion to say that those virtues are less important than God, for the sufficient reason that God anchors them in the nature of things. James Russell Lowell's familiar lines are dedicated to this point:

Truth forever on the scaffold, Wrong
forever on the throne, —
Yet that scaffold sways the future, and,
behind the dim unknown,
Standeth God within the shadow, keep-
ing watch above his own.

I repeat this important point. The issue is not over compassion and an alternative of whatever sort, but over the status of compassion in the nature of things. Is compassion rooted in ultimate reality, or is it only an admirable human virtue? That is a vertical question pertaining to worldviews. Liberals inherited their exemplary passion for social justice from parents and grandparents who (for all their social concern) nailed the horizontal arm of the Christian cross to its vertical arm which (in standard rendition) is longer to symbolize its priority. In their declining concern for theology and worldviews, liberal Christians have in effect turned the cross on its side and made its horizontal arm the longer of the two.

CHAPTER 14

The Big Picture

The cosmic Rorschach inkblot includes everything. And, because backgrounds influence foregrounds, what we attend to focally is influenced by our background *sense* of everything, of the whole. I say "*sense* of the whole" because backgrounds are not in direct view. To become conscious of them we have to redirect our gaze and turn them into foregrounds.

The remainder of this book does that—turns background into foreground with the traditional Big Picture, the background against which human life was lived until the scientific backdrop replaced it. I will remind the reader that I used the opening chapter of this book to suggest that a sensible way to enter the third millennium would be to pass a strainer through the three periods of the human past and carry forward the best in each while leaving the dead to bury the dead with respect to the remainder. The best thing about modernism was its science, the best thing about postmodernism was/is its concern for justice, and the best thing about the traditional age was/is its worldview.

Wanting first to establish the reasons for my interest in the traditional period, I have delayed until this chapter my full description of it. That no two of the over seventy thousand estimated societies that have existed are photocopies of each other poses no problem, for I have devoted an entire book to that matter. Darrol Bryant took the two initial essays that I wrote on the subject—"Accents of the World's Philosophies" and "Accents of the World's Religions"—and added to them my fourteen subsequent essays on the subject to edit the whole into a book he titled *Huston Smith: Essays on World Religion.* Differences having been dealt with there, I am free here to focus on the conceptual spine that underlies those differences. A second earlier book, *Forgotten Truth: The Common Vision of the World's Religions,* describes that spine in detail. Here I will compress the major findings of that book and state them in ways that I hope will be accessible to the general public.

Think of this chapter, if you will, as the "generative grammar" that gave rise to the manifold natural languages of the human spirit, the world's religious outlooks. The language of science is not a natural language. *Lingua franca* of our times though it has become, it is an artificial language that cannot accommodate the human spirit.

The Great Divide

Looking out upon their world, traditional peoples divide it into *this-world* and an *Other-world.* Other animals do not make this division, and early on human beings may not have done so either, for the theme of an original wholeness at the beginning of time—the Garden of Eden story in some form or other—turns up repeatedly. Be that as it may, the earliest human mentality that has survived on our planet—that of the Australian aborigines who did not experience an iron age—shows the this-world/Other-world divide solidly in place.

The aborigines call their Other-world "the Dreaming" and contrast it with the everyday world because of its immunity to time. Things in the ordinary world come and go, but time does not touch the Dreaming. It is peopled with legendary figures who are much like ourselves while at the same time being larger than life. The exceptional status of those mythic figures derives from the fact that they originated life's basic activities. One primordial hero hunted and thereby set that act permanently in place. Another grubbed for roots, another wove baskets, a primordial couple made love and begat children—all of these and more until every basic kind of human act was ensconced.

Outsiders would suppose that when an aborigine engages in a given act he thinks of himself as *imitating* the hero who originated it, but that would be too weak. In early mentality the line between this-world and the Other-world is thin: the aborigine *identifies* with the initiating hero to the point of *becoming* that hero while he is in the Dreaming. And in so doing, he takes on the hero's immortality, for as was just said, time has no purchase on the Dreaming. The goal of aboriginal life is to live as fully as possible in the Dreaming, for that (as the slang expression has it) is "really living." All else is inconsequential.

When we proceed from this earliest this-world/Other-world divide to recorded history, we find the division continuing. Plato's allegory of the cave provided Western civilization with its presiding philosophical metaphor by announcing a stupendous Other (the Sun and its light), compared with which all else is no more than shadows in a cave; and Moses' vision of Mount Sinai in flames added its explicitly religious counterpart. Every religion—and traditional philosophy as well, for traditionally the two were inseparable—turns on this distinction however conceptualized, to the point that its presence can be said to be what makes a worldview religious. Mircea Eliade took this for granted in titling his survey of religious history *The Sacred and the Profane,* and Carlos Castaneda alluded to it when he titled one of his books *A Separate Reality.* In

India the distinction is between *samsara* and *Nirvana.* In East Asia the division is simplicity itself—between earth and Heaven.

Each of the two halves of the traditional worldview then subdivides, which gives us four domains in all. Before turning to the subdivisions, however, I want to stay with this first cut to let its implications (profiled below) sink in:

1. As we saw earlier, the standard terms for designating the two halves of the world are *immanence* and *Transcendence,* with the latter capitalized to indicate its superiority. In the tunnel metaphor of this book, the darkness within the tunnel represents immanence, and Transcendence is the great outdoors through which the tunnel runs.

2. For definitional clarity, I have thus far referred to this-world and the Other-world as if they were halves of an apple, but that is misleading. The truth of the matter is contained in the diagram shown in Figure 1 (see Chapter Twelve), which has the physical universe as the smaller circle within a larger one that includes it and more besides. Situating it thus allows the attributes of the larger circle to wash through the smaller one—the circumference of the smaller circle is perforated to allow for this—but religious sensibilities are required to detect those incursions, for they are hidden within nature's outward appearances. Science has no reason to take account of this point, but it is vital for religion. God's omnipresence announces the point abstractly, but concrete expressions are more telling, so I shall enter four.

When the Buddha awakened under the bo tree, his first exclamation was, "Wonder of wonders; all things intrinsically *are* the Buddha-nature." The refrain that resounds through *The Heart Sutra* like a rhythmic gong-beat is: "Form is emptiness, emptiness is form; there is no form without emptiness, there is no emptiness without form." The psalmist proclaims that "heaven and earth are full of Thy glory." Finally, St. Paul assures us that "in Him we live, and move, and have our being."

These testaments bring out the limitations of spatial metaphors in dealing with Spirit. The distinction between this-world and the Other-world is more accurately understood as a matter of perception, not geography. What are we able to *see* with what Plato called the eye of the soul and Sufis refer to as the eye of the heart? In Blake's formulation, "If the doors of perception were cleansed, we would see everything as it is: infinite."

3. Metaphysically speaking, there is no clearer way of characterizing modernism and postmodernism than to say that their world consists of this-world only. I repeat for a final time the telling line from the *Chronicle of Higher Education*: "If anything characterizes 'modernity,' it is a loss of faith in transcendence, in a reality that encompasses but surpasses our quotidian affairs."

4. Having learned as a teacher that repetition never hurts, I add as my fourth point here what this book has been arguing throughout. We have dropped Transcendence not because we have discovered something that proves it nonexistent. We have merely lowered our gaze. The toll this has exacted occupied the first half of this book.

5. All the while, science preaches two worlds of its own that parallel those of religion with the difference that the other-world of science is quantitative whereas that of religion is qualitative. Like religion's Other-world, the quantum world is invisible to human eyes while determining what these eyes perceive. Again like the Other-world, the quantum world is not elsewhere—one burrows inward to find it. And still again, the quantum world too is strange to the point of being barely intelligible.

Subdivisions

Coming now to the subdivisions in the two halves of the Big Picture, this-world divides into its visible and its invisible components, and the Other-world into its knowable and ineffable aspects. I begin with this-world.

The Two Halves of This-World

Prior to the invention of the magnifying glass, the visible world consisted of what our physical senses report, but modes of amplification have extended our senses deeper into nature. This makes it better to think of the visible world as the physical universe in its entirety—which is to say, what we pick up with our unaided senses plus what science adds to their reports.

Turning to the invisible, immaterial half of this-world, we encounter it directly in our thoughts and feelings, but the traditional and modern views differ categorically as to how far into nature immaterial things extend. Traditionalists consider discarnates—angels, demons, patron saints, shamanic allies, and their likes—to be as much a part of the world's furniture as are mountains and rivers, but modernity has withdrawn consciousness (or more broadly, *sentience*) from the world at large by making it an epiphenomenon of biological organisms at some level of their complexity. Since life exists only on our planet (it seems), the contraction is almost total. Sentience exists only on our planetary mote in the sidereal universe—a mote so small that it barely escapes being a mathematical point—and only in the rivulets of life on this mote. If a large asteroid were to destroy planet earth, the universe would consist of nothing but dead matter.

My object in this chapter is to describe the Big Picture, not to argue for it, but as someone who was raised in a traditional culture and has spent most of his career teaching the Big Picture as impounded in the world's great religions, I find modernity's withdrawal of sentience from the world at large so arbitrary that I shall sandwich in the single firsthand experience I have had that tugs against it.

It was 1957, and I was serving as visiting professor at Stephens College in Columbia, Missouri. My television course on world religions had premiered in St. Louis the preceding spring, and that made me a bit of a celebrity in Columbia that semester. So it was that I was invited to meet John Neihardt, author of *Black Elk Speaks,* who was the literary showpiece at the University of Missouri. To prepare for the meeting I reread his book and arrived at his house prepared to discuss Black Elk, only to find him so obsessed with a recent event that his book scarcely entered our conversation. The story that poured from him and his wife was this.

The preceding week the Neihardts had been involved in a minor automobile accident. Nothing serious—scraped fenders, a few dents, something like that—but in those days local insurance agents paid house calls to gather details. The Neihardts had been sitting at the dining-room table explaining the accident to the agent when he interrupted and said, "Would you mind putting your dog out? It's making me nervous."

"Dog? What dog?" the Neihardts wanted to know.

"Oh, you know. That little black spaniel." He glanced under the table and, seeing nothing, added, "He must have gone out."

The Neihardts looked at each other in astonishment. They had had a black spaniel who had been the joy of their lives, but he had died of old age the week before.

That was where the story from them ended, but years later I learned of its conclusion. I referred to the incident in a newspaper interview that came to the eyes of a couple who had been personal friends of the Neihardts, and they wrote me to fill me in on the story's end. The Neihardts (who were close to retirement when this incident occurred) devoted the rest of their lives to studying parapsychology and bequeathed their estate to the founding of a research center for studying the paranormal.

That small assist to the traditional understanding of this-world is not conclusive. It does not require the conclusion that the spaniel's soul continued after he died and continued to impact the living, for it is equally possible that the insurance agent picked up telepathically on the Neihardts' remembrances of their dog. Still, telepathy too is not a part of the standard scientific worldview, so in either case the reported facts seem to challenge that view in one way or another. I will leave matters there and proceed to the Other-world.

The Two Halves of the Other-World

Everywhere it subdivides into God's knowable aspects, on the one hand, and on the other, God's unfathomable depths, which Jacob Boehme called the Divine Abyss and Meister Eckhart christened the Godhead. (Asian equivalents will be entered in due course.) The distinction can also be described philosophically (as Neoplatonists and Vedantists do), but I will limit myself here to the theistic expressions. I shall use *God* and *Godhead* as my generic names for the division in the Other-world, but there are other pairs of terms that help in understanding the division. I shall take note of three: God as knowable and unknowable, God as manifest and hidden, and God as personal and transpersonal.

God as knowable and unknowable. Insofar as *unknowable* denotes ignorance, it is misleading here, for we are not totally ignorant of the Godhead. Only left-brain conceptual knowledge—knowledge that can be put into words—is decommissioned. The Godhead cannot be rationally described, but (in a way that resembles seeing more than thinking) it can be intuited—or better, *intuitively discerned.* Job's climactic testament is paradigmatic here: "I had heard of thee by the hearing of the ear, but now my eye sees thee." We sense the "seeing" too in the word I have set in Roman in this untitled poem by Eunice Tietjens, which appears in a book of photographs titled *Everest: The West Ridge:*

The stone grows old,
Eternity is not for stones.
But I shall go down from this airy space, this swift white
peace, this stinging exultation;
And time will close about me, and my soul stir to the
rhythm of the daily round.
Yet, having known, *life will not press so close.*
And always I shall feel time ravel thin about me.
For once I stood
In the white windy presence of eternity.

Manifest and hidden. These are two of the Ninety-nine Beautiful Names of Allah, *az-zahir* and *al-batin.* Human beings resemble God in likewise having hidden and manifest aspects. Our physical features are open to the world, while not even our friends and relatives are privy to our inner lives in their inscrutable depths.

Personal and transpersonal. Montesquieu quipped that if triangles had gods, those gods would have three sides. He intended his remark as satire, but it contains an important truth. We best understand things that resemble us. Enter the personal God, decked out in attributes like ours (though they exceed ours infinitely in nobility). It seems altogether reasonable that from among God's infinite attributes human sensibilities would most readily discern virtues that they themselves possess (goodness, mercy, love, and their likes).

The logical complement to the personal God is the transpersonal God, but we must be careful here. The obvious mistake would be to confuse *transpersonal* with *impersonal,* which of course God cannot be. Transpersonal is *more* than personal, not less, and that makes it a difficult notion: it is not easy to imagine things that outdistance us. If one restricts *personal* to meaning "having a center of self-awareness," it is applicable, because God is never without that; but the primary meaning of the word derives from human beings, *human* persons, and as such persons are radically finite, *personal* is a tricky adjective to attach to God. People who have trouble with the notion of a personal God—their number seems to be increasing—are put off because the concept cloys for sounding anthropomorphic. (That is how Spinoza and his disciple Einstein saw the matter.) They have a point. To be religiously available, God must resemble us in some ways or we could not relate to him. Yet too much like us, God ceases to evoke the reverence and awe that are required for worship. Likeness and difference—both are required; and at their best, they work together in counterpoint. That there are not two Gods goes without saying. We are talking about degrees in understanding a single reality.

Having delineated *God* from *Godhead,* it remains only to indicate the ubiquity of the distinction.

Making due allowance not only for differences in terminology but for differences in nuances, in East Asia we find Confucianism's *shang ti,* the supreme ancestor, and beyond him *Tien,* or Heaven. In Taoism, there is the *tao* that can be spoken, and the *Tao* that transcends speech.

In South Asia, Hinduism presents us with *saguna brahman*—God with attributes or qualities, among which *sat, chit,* and *ananda* (infinite being, awareness, and bliss) are primary—and *Nirguna Brahman,* the *neti, neti* (not this, not this) of the Brahman who is beyond all qualities. Buddhism presents a special case because of its ambiguous stance toward God, but though the personal God is absent in early Buddhism, it could not be excluded indefinitely and came pouring in through the *Mahayana.* A recent newspaper article quotes the abbot of a temple in southern California as saying to his morning congregation, "In the morning you feel your heart touch the Buddha's heart, and Buddha's heart is so happy, so full of kindness." The transpersonal God is, of course, solidly ensconced in Buddhism's *sunyata*—emptiness—and *Nirvana.*

Finally (and reversing the situation in Buddhism), the Western, Abrahamic family of religions pulls out the stops on the personal God—the God of Abraham, Isaac, and Jacob, the Father of our Lord and Savior Jesus Christ, and Allah of the Ninety-nine Beautiful Names. Still, the transpersonal God is not lacking. In Judaism we glimpse it in the *'ein sof* (Infinite) of the Kabbalah. In Christianity it shows itself in *The Cloud of Unknowing*, Meister Eckhart's Godhead, and Paul Tillich's God-beyond-God. And in Islam it is Allah's Hundredth Beautiful Name, which (because it is unutterable) is absent from the Sufi rosary.

The human self is the only creature that in its completeness intersects all four regions of reality that have now been demarcated. In a transfixing moment that I shall never forget, I once saw Crater Lake backed by mountains that were topped by a cumulus of white clouds before the empyrean blue sky took over above them. The lake was so glassy that what I saw rising above it I also saw inverted, in a perfect mirror image, in the lake's depths. It struck me as an analogue for the way the levels of reality are mirrored in the human self, where they too are inverted imagistically. As a mandala to end all mandalas, I offer in Figure 2 a diagram that shows how the four regions of reality (and of selfhood, I have now added) turn up cross-culturally.

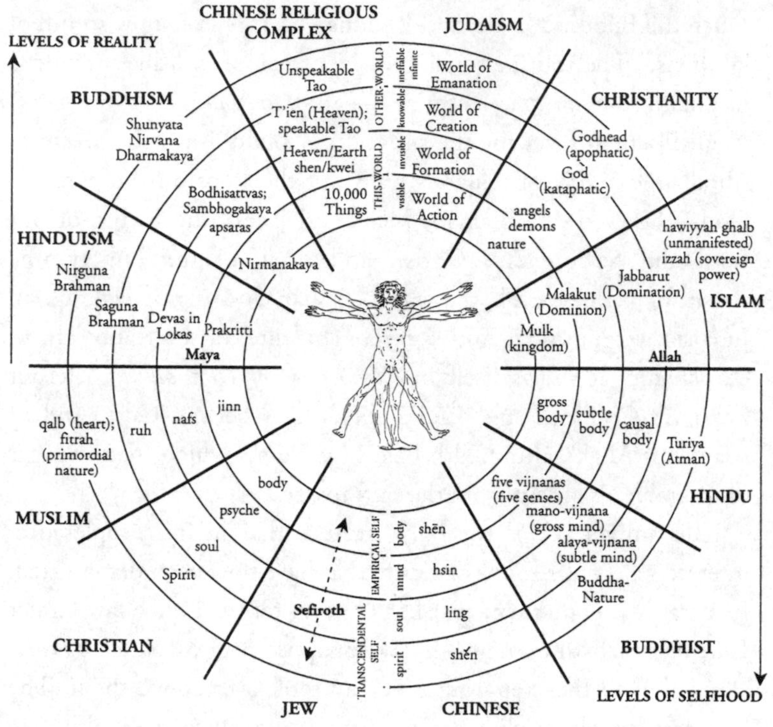

Figure 2.

Graphic layout courtesy of Brad Reynolds.

A Hierarchical Reality

With the four regions of reality elaborated, I turn now to their relationships. The central point has been implied throughout this chapter, but it needs to be stated explicitly. The four domains are not identical in worth. We caught sight of this in our glance at the aboriginal Dreaming, which is incomparably more worth-filled than is mundane existence, and when we fan out the world into its four regions, they together present us with a hierarchical worldview. Being infinite, the Godhead is more complete than God, who in turn is more important than the two halves of this-world (which are better considered together here, for neither is clearly superior to the other). Unfortunately, the word *hierarchy* has fallen on troubled days, which requires that I take a paragraph to rehabilitate it before proceeding.

Etymologically, *hierarchy* comes close to being a perfect word for joining the two virtues—holiness, *heiros,* and sovereign power, *arkhes*—which, conjoined, announce religion's central claim. For as William James stated that claim, "Religion says that the best things are the more eternal things, the things in the universe that throw the last stone, so to speak, and say the final word." However, reckless assaults on the word from what Frederick Crews has called the eclectic left have all but ruined it by building oppression into its very definition. By definitional fiat, this turns "empowering hierarchy" into an oxymoron and leaves the general public without a word for chains of command that are legitimate and enabling. That there are such chains is obvious once the matter is given a moment's thought, however. A loving family with small children is an empowering hierarchy, as is a well-run classroom. The definitive example of a benevolent hierarchy is God's relation to the world, which Christians compress into the formula that was mentioned in Chapter Two: "God became man that man might become God."

That said, I now proceed to the traditional hierarchical worldview. Every virtue increases as we mount from this-world (its two halves taken together) through God to the Godhead, where they reach their logical limits. We cannot concretely imagine those limits—perfection, omniscience, omnipotence, omnipresence, and their likes are beyond our understanding—but we *can* follow the logic of the matter, and in any case we do know what the virtues are through the rudimentary ways they surface in us. Ones that come immediately to mind are the Greek ternary of the good, the true, and the beautiful; India's already mentioned existence, consciousness, and bliss; the creativity and compassion that Yahweh so steadfastly exemplifies; and in their full sweep, Islam's Ninety-nine Beautiful Names of Allah. Christian love should not be overlooked, nor should the power that in God climaxes in omnipotence. The diagram that follows (Figure 3) presents the matter graphically.

In us the virtues are distinct—knowledge is not the same as beauty, and neither of those is synonymous with power. In God the virtues overlap while remaining distinguishable, but in the mathematical point of the Godhead at the top of the diagram, the boundaries between the virtues dissolve and each takes on the features of the others. The Godhead knows lovingly and loves knowingly, and so on, until all the virtues smelt down into a singularity that the Scholastics called "the divine simplicity."

Topdown Causation and the Multiple Degrees of Reality

In direct opposition to the scientific worldview, where causation proceeds upward from simple to complex, in the traditional worldview causation is downward, from superior to inferior. Whether one speaks of God, or of Amaterasu (the Heavenly-Shining Deity

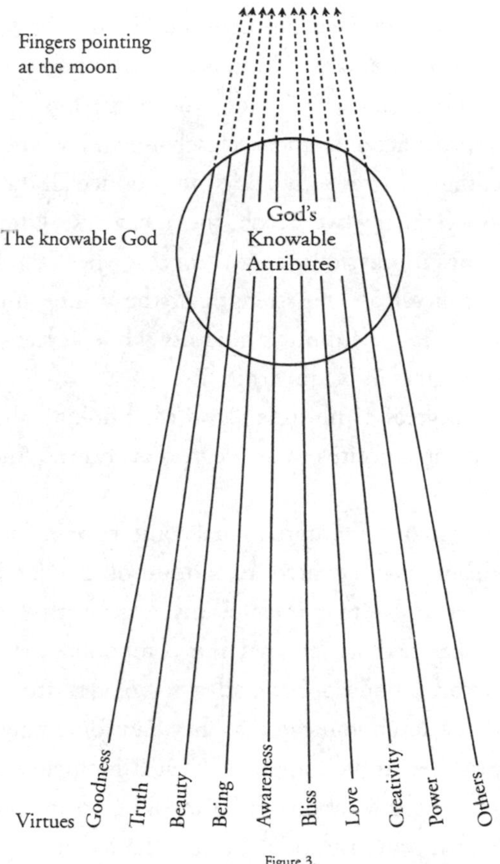

Figure 3.

of Japan's creation myth) creating the world, or of its emanating from the Transpersonal One (as Neoplatonists, Vedantists, and philosophical Taoists prefer to do), effects never equal their causes.

A frequent metaphor invoked to convey this idea is *veiling*. The Infinite cannot (on pain of self-contradiction) renounce its infinity. Yet at the same time, *because* it is infinite, it can exclude nothing, which means that it must include finitude. By the same line of reasoning, not only must it include finitude, it must include all of its gradations. (Speculating, Freeman Dyson calls this "the principle of maximum diversity," which would make the universe as interesting as possible.) Only a single veil hides the fullness of the Infinite from what approximates it most closely—namely, the personal God—but veils are progressively added to produce all the grades of finitude until eventually we reach the meagerest kind of existence—the strings in physics' string theory?—where the Infinite is almost totally concealed. The strength of the veiling metaphor is that it acknowledges the Infinite's ubiquity while at the same time accounting for degrees in its discernibility.

The phrase "degrees of finitude" is worth thinking about, especially when it is stated positively, as "degrees of reality." And thereon hangs a tale.

A decade or so ago at an international conference in Seoul, South Korea, participants were treated to a tour of the famed Palace Gardens. I regret to say that their beauty was largely lost on me, because no sooner had we entered the compound than I found myself flanked by a philosopher and a sociologist from England whose determined air announced that they had something in mind. So resolute were they in their approach that it momentarily crossed my mind that I might be about to be whisked off for purposes of interrogation. That proved to be the case as far as interrogation was concerned, though not the whisked-off part. Somewhere in one of my books they had come upon a discussion of degrees of reality, and they were bent on finding out if I really considered the phrase to be meaningful. How could a state of affairs be anything other than simply a state of affairs? they wanted to know.

As the reader is gathering, I found the experience rather unnerving. For starters, it was two against one, besides which Oxford English makes me feel inferior before ideas even enter. I will, therefore, characterize the outcome minimally by saying that our "discussion" did not end in a smashing victory for degrees of reality.

Still, win one, lose one—or, in this case, lose one, win one. There is a sequel to the Korea story that has me doing better.

While I was at Syracuse University the religion department had a practice of mounting departmental colloquia to celebrate the publication of books by its faculty. Three colleagues would comment on the book and a discussion followed. When my *Forgotten Truth* appeared, it fell to my lot to have my departmental chairman as one of the three respondents.

I cannot remember anything else that was said on the afternoon in question, but I remember as if it were yesterday my chairman's report of his reaction to my book. He confessed that departmental duties had forced him to put off reading it until the last minute, but he had blocked off the preceding evening and the morning of the colloquium to prepare. As things turned out, he needed the morning only to catch up on his sleep, for by the end of the book's first chapter he had found himself so agitated that he read the book straight through (which meant getting to bed around two in the morning). Never had he encountered such a thing, he reported: a respected scholar defending the traditional—*hierarchical,* no less!—worldview as being superior to the science-backed metaphysics of modernity. What baffled him most turned out to be the same point that my British friends had cornered me on—the notion that the hierarchy in question turned on degrees of reality. What could that possibly mean?

I have no memory of how that afternoon ended except that we both phoned our wives to say we would not be home for dinner and hied ourselves to a nearby restaurant to continue the "discussion," to put what followed politely. Degrees of reality remained central, and I come now to my reason for telling this second story.

We were making no progress and had descended into recycling our arguments when suddenly my chairman grew quiet. I assumed that the pause was to attend to his dinner, which was growing cold, but I was mistaken. After what seemed like an exceptionally long wait, he began to speak again, this time in a different tone. He had remembered something that had happened the previous evening before he settled into reading my book. Entering his living room, he had found his six-year-old son watching television, and the boy was agitated. With reason, for violence was everywhere; people were getting killed right and left. In a voice that bordered on genuine fright, his son had turned to him and asked, "Dad, is this real?"

Out of the mouths of babes. These are curious times, when a six-year-old can understand that "real" has degrees, while his philosophically trained father cannot. Add hucksters to the son's company, for a cereal box that recently caught my eye assured me that it contained "the real thing." And ordinary speech gets into the act as well. At one time or another we have all found ourselves exiting a sports event exclaiming to one another, "That was a *real* game." A student once told me that he never registered for courses in advance because he spent the first week of the semester shopping around to determine which professors were, as he put it, "for real." (I flatter myself in thinking that because he did register for my course I passed his test.)

I would not have devoted this much space to this reality issue if it did not come close to being the heart of the contest between the traditional and the modern worldviews. In a single-storied world where there is no Transcendence, capitalizing *Reality* does not give the word a distinctive referent; all it does is to pump enthusiasm into the word. As one British analytic philosopher (I have forgotten his name) put the matter, Reality when capitalized means nothing more than "reality, loud cheers."

I am having trouble putting this particular issue behind me, but this final entry should do the trick. My church has picked up the idea of building advertising into its return address, so in the upper lefthand corner of its envelopes the name of the church is now followed by "Committed to Social Justice and Spiritual Growth." Fancifully (but only partly so) I have found myself playing with an alternative wording that would read, "Committed to Making People Real," for that is not a bad way of describing the religious project: the effort to transcend phoniness. The whole object of religion, it might be said, is to enable people to come as close as possible to God's infinite reality. That should be easy, because God is so real that we should respond like iron filings to his magnetic pull. Actually, though, it is difficult, because we are so *unreal* that there is not much in us for God's pull to grab hold of. Would it not be refreshing to learn from a church's return address that it was "Committed to Making People Less Shallow"?

Return to the Inkblot

Stating matters in the simplest way I can manage, I have presented the conceptual spine of the traditional worldview. To modern ears it is likely to sound archaic if not arcane, as when we find E. O. Wilson saying in his *Consilience* that prescientific opinions about the world are "wrong, always wrong." But I have already stated the decisive point, which is that science has discovered nothing in the way of objective facts that counts against traditional metaphysics.

Wilson cites no such facts because he does not understand the difference between cosmology (where his assertion holds) and metaphysics (where it does not). The traditional worldview can incorporate everything that science has discovered without a ripple, for it fits comfortably into the smaller circle of Figure 1 (chapter 12) that is contained in the larger circle. This leaves no more than styles of thought barring us from looking again at the traditional worldview, and styles come and go.

Wilson's quoted assertion is wrong, but that does not make the traditional worldview right. There will be no backing away from the conclusion of the preceding chapter, which was that worldviews are unprovable. There are, however, ideas that are worth pondering as we decide which outlook we want to live by, and I shall mention three.

1. Can something derive from nothing? Can a stream rise higher than its source? Intuitively, neither seems likely, but the scientific view requires affirmative answers to the questions whereas the traditional worldview does not. Life from non-life, sentience from insentience, intelligence from what lacks it—for science it is more-deriving-from-less at every step.

2. Three chapters back I cited Colin McGinn's *The Mysterious Flame,* and I want to ask again whether there are any reasons against thinking that human beings have *three* innate talents, a talent for the Big Picture, as well as for language and science. If the multiplicity of such Pictures—the variations in religion from culture to culture—is put forward as counting against the second talent, I will repeat what I have said several times: that the multiplicity comes through to us as variations on a common theme. For as Ken Wilber has written, the concept of a hierarchical worldview (along the lines I have outlined in this chapter) is "so overwhelmingly widespread that it is either the single greatest intellectual error ever

to appear in human history—an error so colossally widespread as to literally stagger the mind—or it is the most accurate reflection of reality to have appeared."

To this I will myself add that at minimum this view of reality emerges as the view that conforms most closely to the full range of human intuitions. As I put the matter in my *Forgotten Truth:*

> Consituting until recently, through both rumored and recorded history, what we have ventured to call the human unanimity—the phrase overstates the case slightly, but not much—it presents itself as the natural human outlook, the view that is normal to the human station because it is consonant with the complete complement of human sensibilities. It is the visions philosophers have dreamed, mystics have seen, and prophets have enunciated.

3. We fully understand physical objects such as cars when we can ourselves build them. Elsewhere "explanation" becomes a tricky business. Philosophers have found no criterion for when something can be said to have been explained then that the proffered explanation satisfies us. To this, Archbishop and theologian William Temple adds that the sought-for satisfaction arrives when the explanation shows that what is being explained is found to be as we think it should be. If someone sets out to build a better mousetrap, his objective makes perfect sense and all that remains is for us to wish him luck. But if someone set out to build a worse mousetrap, *that* might require the explanations of a psychiatrist. Transposed to metaphysics, the moral is this: true or not, the traditional worldview is transparently intelligible. The scientific worldview is not. Final causes being categorically excluded from it, it necessarily deadends in questions that have no answer.

CHAPTER 15

SPIRITUAL PERSONALITY TYPES

I used to think that the most important religious differences are those between the great historical religions—in our day, Hinduism, Buddhism, Judaism, Christianity, Islam, and their likes (including tribal religions among Native Americans and others). Increasingly, however, I have become convinced that there is a deeper set of differences that cuts across these institutional lines. In every sizable community one finds atheists who think that there is no God, polytheists who acknowledge many gods, monotheists who believe in a single God, and mystics who say that there is only God.

These four ways of slicing the religious pie (if this expression may be excused) are not explicitly articulated in the way theologies are. For the most part they pass unnoticed, for they leave no footprints in history and do not create headlines, as religions do when they collide. Yet the differences between the four spiritual personality types (as I am calling them) run deeper than theological differences, for they are grounded in human nature, whereas theological differences, being historical, come and go.

What demarcates the four types is the size of the world each inhabits. Beginning with the smallest, the atheists' world houses nothing but matter and the subjective experiences of biological organisms. Polytheists add spirits to the foregoing—this is the realm of folk religion, which is much the same the world over. Monotheists place all of the above under the aegis of a supreme being who creates and orchestrates everything. Having nothing further to add to the foregoing, mystics double back over the terrain to find God everywhere.

This way of putting things seems to give each successive type the advantage over its predecessors for having a more commodious world to live in, but everything turns on whether the larger worlds exist. For Atheists, they do not; the larger worlds are seen as projections of the human imagination. And the same holds at the other way-stations. For Polytheists, for example, the idea of a single, all-accountable God may be assented to, but it has little direct bearing on the lives they actually live.

If we visualize the four worlds in the way they are diagrammed in Figure 2, Chapter Fourteen, and mount a vertical axis that extends from the circle's center upward, we can think of the lines that separate the four levels of reality as one-way mirrors. For someone looking upward from the center, the lines are mirrors. One sees nothing above them; what one sees in looking at them is reflections of things on one's own plane. From the other side, however, they are plate glass. Things on the levels below the glass are in plain view.

This analogy will be developed in later pages, but here at the chapter's start let me say that it is designed to produce an impartial way of seeing the four personality types in relation to one another. Each type can argue that the world ends where its cosmic mirror lies and that those who posit things beyond that point are merely projecting in the psychological meaning of that word. (This ties in with the contention of Chapter Thirteen that the world is religiously ambiguous.) Before expanding on this metaphor, however, I shall first situate this chapter in the context of a perennial human preoccupation: characterology.

CHARACTEROLOGY

Know your type, we are told, and innumerable people give large chunks of their lives to trying to do just that. In the ritual of the Sunday morning newspaper, readers who turn first to their horoscopes compete with those who glance first at the comic strips or financial pages. Add to this the interest in the four Jungian types (thinking, feeling, intuitive, and sensing)—often identified with the help of the Myers-Briggs type indicator—and the nine-fold Enneagram and it is clear that we are into a lively topic.

It has deep roots. Astrology is universal and goes back as far as we can see, with numerous cultural overlays. Psychologically obsessed India has not one but three complementing ways of classifying people—by their *yoga* (most effective way of approaching God), their *varnas* (social stations), and their *gunas* (predominant psychological dispositions). The West's longest-standing classification, which dates from Empedocles, Hippocrates, and Galen, links four basic human temperaments (sanguine, phlegmatic, choleric, and melancholic) to their respective natural elements (air, water, fire, and earth) and bodily humors (blood, phlegm, yellow bile, and black bile). We still speak of unemotional persons as phlegmatic (phlegm-ridden); cheerful, unflappable ones as sanguine (blood-dominated); hot-headed ones as choleric (from the Greek *khole,* for bile); and sullen, depressed ones as melancholic (from *melas,* black, and again *khole,* bile).

My object in this chapter is to add to the world's storehouse of typologies one that takes the human spirit as its chief concern. The next and final chapter of this book will argue that Spirit is what ties us directly to the Big Picture, and spiritual personality types are determined by the size of their respective Big Picture. The preceding chapter presented the four major Pictures that people have drawn from the cosmic Rorschach blot. The four spiritual personality types are defined by which of these worlds each type believes exists.

Ubiquity

The four types turn up not only everywhere but everywhen, for we can track them as far back as historians can see. Instead of cataloguing this claim (which would prove tedious), I will simply do a headcount in places where we might expect one of them to be missing. In discussing the ways in which the cosmic inkblot can be read, I have already pointed out that in tribal societies, where one might expect to find uniformity, the village atheist still turns up. At the other end of the temporal spectrum, we might assume that modernism considers polytheism superstitious, but not so, as the following anecdote reveals.

When Robert Graves was at MIT for three weeks, departments were invited to host him for dinners, and we philosophers gladly availed ourselves of that opportunity. When the table had been cleared for after-dinner drinks, Graves lit a cigar, leaned back in his chair, and, addressing us squarely, asked, "What do you gentlemen have against ghosts?" I thought that our chairman, Hilary Putnam, would choke on his brandy before he recovered himself and turned the conversation toward Graves's love poetry.

It was surprising to find disembodied spirits turning up in an MIT faculty gathering—students are a different population—but I have already pointed out that in the New Age Movement gods and goddesses are everywhere. Monotheism is officially absent from early and southern Buddhism—I say *officially,* because I have had taxi drivers in Sri Lanka stop before commencing long drives to light josh-sticks before images of the Buddha—but it enters torrentially in the *Mahayana.* (This too has been previously noted.) In my graduate school days I had to *hunt* for mysticism in the severely monotheistic Abrahamic religions, but today those who otherwise have no use for religion generally make an exception for mystics. One thinks of the respect accorded to Jalal ad-Din al-Rumi. In passing, we all have all four types within us; the differences are in degree.

Spot-check completed, I proceed now to the types themselves.

The Atheist: There Is No God

As the label indicates, atheism is a negative stance. (In Greek, the prefix *a* announces negation: the gnostic knows; the agnostic does not know.) It is important to make clear that in the present typology the negativity pertains only to disbelief in worldviews that house God. It has nothing to do with the atheist's stance toward life (which is as likely to be as life-affirming as the next man's), or with character traits of any sort. This needs to be clearly stated, because mention of the word *atheism* draws *ad hominem* judgments, both positive and negative, in the way carcasses draw flies. During the McCarthy era, when the Cold War was at its height, atheism was associated so strongly with Communism that "atheistic Communism" began to sound like a single word. Today something of the reverse pertains. The bad name that modernity has given religion has (in the knowledge culture) moved virtues from the theistic to the atheistic side of the ledger. To support his atheism, Albert Camus told us that early in his life he resolved to live without lying, the implication being that religious people live by lies. In the same vein, Einstein said that "in their struggle for the ethical good, teachers must have the stature to give up the personal God."

A major virtue of my typology is that it steers clear of *ad hominems* of all varieties to attend exclusively to the Big Pictures that personality types subscribe to. The atheist's world stops with the inner circle in Figure 2, Chapter 14, with sentience added as an epiphenomenon of biological organisms. The physical universe as conceived by science and common sense is what exists. This comes to fifteen billion light-years of dead matter with the subjective experiences of biological organisms appended.

THE POLYTHEIST: THERE ARE MANY GODS

The Polytheist's world consists of this-world in its traditional fullness. In that fullness, gods, spirits, and discarnates are taken for granted as much as chairs and tables. They are fully as real, and for all we know are equally numerous. As recently as a century or two ago, angels, demons, and patron saints were thought to be everywhere, actively engaged in the human drama. Today when we come upon warlocks and witches, nymphs and woodland sprites, goblins, ghosts, and the elves and "little people" of the Irish, we assume that we are into folklore. Earlier, such things were at the heart of folk *religion*. Shamans dealt with spirits directly, as did mediums with their occult guides and "controls." Belief in these spirits was in no wise confined to lower classes. In the eighteenth and nineteenth centuries the Russian royal court was honeycombed with spiritualism; much earlier, Socrates had his *daemon,* who advised him what not to do (though never what he should do).

Spirits are not necessarily good. In the Chinese town where I grew up, evil spirits predominated; indeed, warding them off seemed to be the town's operative religion. Bottles over doorways (their nozzles pointed outward to fool spirits into thinking that they were cannons), disappeared with the science-addicted Communist takeover, but curiously the town's main monument to evil spirits still survives. This story is this:

In the nineteenth century a plague ravaged the town, and geomancers determined that it was being caused by evil spirits that were pouring into town through the city's west gate. The town fathers responded by building a squat, airtight pagoda near the gate. Before its final wall was completed, they commissioned Taoist priests to lure the spirits into the structure, where (with the final wall completed) they presumably still remain. I find it interesting that with all the Communists' talk about eradicating superstition, they have let that structure stand. It raises the suspicion that somewhere in the recesses of the Chinese psyche there remains a touch of the polytheist that official Communism has not eradicated. You never know. Better safe than sorry. It's wise to let well enough alone.

Spirits are normally invisible. Angels do show themselves occasionally—in the Abrahamic tradition, Gabriel displays that talent repeatedly—but the talent itself is unusual. Invisibility usually implies immateriality, but not entirely here, for references to spirits having bodies are widespread. To cite but two examples: in Mahayana Buddhism we have the three bodies *(kayas)* of the Buddha, only one of which was protoplasmic (the body of Siddhartha Gautama in his earthly incarnations); and in Christianity we have Christ's glorified body, which—following his resurrection—slid through closed doors before it ascended to heaven.

What this comes down to is that discarnates are exempt from *gross* matter and the matrices of space, time, and matter that govern it. Ghosts glide through walls in the way lasers move through lead plates, and angels are said to change locations simply by *willing* to do so. Presumably the Taoist priests who imprisoned the evil spirits in my hometown stripped them of this talent—but enough; logic is useless in the spirit world. One fact, however, comes through unequivocally. We cannot assume that as a class discarnates see things more clearly than we do, or that they are happier and more benevolent. One thinks of the hungry ghosts and hell-beings in Buddhism's Six Realms of Existence.

With the polytheist's world thus sketchily described, what happens when it collides with science? Very little, actually. There is a saying that when the devil conquers the citadel, he seldom bothers to change the flag, and the saying fits quite well here. Modernists no longer turn to spirits for healings and rainfall, now that technology has proved itself more dependable on those fronts. But that does not mean the death of polytheism, for there is more to the polytheist than interest in magical technology.

At root, what makes people polytheists is their refusal to settle for the obvious. Stated positively, polytheists have an irrepressible outreach for something more than mundane existence. When they no longer need the "more" to help with their physical needs, they veer toward the psychic. "People want to marvel at something," the master magician Magnus Eisengrim observes in one of Robinson Davies's novels, "and the whole spirit of our times is not to let them do it. We have educated ourselves into a world from which wonder, and the fear and dread and splendor and freedom to wonder, have been banished."

Polytheists protest that banishment. Consider the following:

- Fascination with the preternatural continues unabated today, which causes even those who reject spirits categorically to concede that interest in them is probably ingrained in the human makeup. Everyone loves a spook yarn—a good ghost story or gothic tale that sends chills down the spine. Science fiction is only the latest chapter in this genre.
- For a sizable proportion of the population, even in the modern West, this fascination phases subtly into belief. How many of our associates who present themselves in the workaday world as no-nonsense meat-and-potatoes types have

(hidden well out of sight) firsthand accounts of a close brush or two with the preternatural? Catch them off-guard and they too have their stories.

- One of the appeals of Jungianism is that it allows people to indulge their polytheistic proclivities while remaining culturally respectable. It accomplishes this by transplanting gods and goddesses from the external world into the collective unconscious. My Jungian colleague at Syracuse University, David Miller, put this viewpoint into a book he titled *The New Polytheism*.
- Polytheists are found within institutionalized churches (whose theology is almost invariably monotheistic) as well as outside them. Sociological studies of religion in the small towns of southern Italy provide us with a typical example. The people of these towns take it for granted that they are good Catholics, and yet operatively, in what matters to them, their Catholicism revolves more around icons and the local patron saint than around the triune God, who seems remote by comparison. What in the end divides the polytheist from the monotheist is not where she stands respecting religious institutions; it is (as I have already said) a temperamental difference, which this time I am approaching from a different angle. The polytheist is interested in the supernatural not for its own sake but for its involvements with this world. And her mind works more concretely. If the invisible is to make a difference to her life, it must impact it palpably.

I will enter an example of this—a single example, both because it is long and because it is vivid enough to make the point by itself. It is from Michael Ondaatje's *The English Patient.* The time is World War II, and inch by inch the Allies are liberating Italy. For those who are not familiar with the word *sapper*—I was not until I came upon this book—it refers to an advance scout charged with the dangerous job of detecting and dismantling land mines.

> When the Eighth Army got to Gabicce on the east coast, the sapper was head of night patrol. On the second night he received a signal over the shortwave that there was enemy movement in the water. The patrol sent out a shell and the water erupted, a rough warning shot. They did not hit anything, but in the white spray of the explosion he picked up a dark outline of movement. He raised the rifle and held the drifting shadow in his sights for a full minute, deciding not to shoot in order to see if there would be another movement nearby. The enemy was still camped up north, in Rimini, on the edge of the city. He had the shadow in his sights when the halo was suddenly illuminated around the head of the Virgin Mary. She was coming out of the sea.
>
> She was standing in a boat. Two men rowed. Two other men held her upright, and as they touched the beach the people of the town began to applaud from their dark and opened widows.
>
> The sapper could see the cream-colored face and the halo of small battery lights. He was lying on the concrete pillbox, between the town and the sea, watching her as the four men climbed out of the boat and lifted the five-foot-tall plaster statue into their arms. They walked up the beach, without pausing, no hesitation for the mines. Perhaps they had watched them being buried and charted them when the Germans had been there. Their feet sank into the sand. This was Gabicce Mare on May 29, 1944. Marine Festival of the Virgin Mary.
>
> Adults and children were on the streets. Men in band uniforms had also emerged. The band would not play and break the rules of curfew, but the instruments were still part of the ceremony, immaculately polished.
>
> He slid from the darkness, the mortar tube strapped to his back, carrying the rifle in his hands. In his turban [he was a

Sikh] and with the weapons he was a shock to them. They did not expect him to emerge too out of the no-man's land of the beach.

He raised his rifle and picked up her face in the gun sight—ageless, without sexuality, the foreground of the men's dark hands reaching into her light, the gracious nod of the twenty small light bulbs. The figure wore a pale blue cloak, her left knee raised slightly to suggest drapery.

These were not romantic people. They had survived the Fascists, the English, Gauls, Goths and Germans. They had been owned so often it meant nothing. But this blue and cream plaster figure had come out of the sea, was placed in a grape truck full of flowers, while the band marched ahead of her in silence. Whatever protection he was supposed to provide for the town was meaningless. He couldn't walk among their children in white dresses with these guns.

He moved one street south of them and walked at the speed of the statue's movement, so they reached the joining streets at the same time. He raised his rifle to pick up her face once again in his sights. It all ended on a promontory overlooking the sea, where they left her and returned to their homes. None of them was aware of his continued presence on the periphery.

Her face was still lit. The four men who had brought her by boat sat in a square around her like sentries. The battery attached to her back began to fade; it died at about four-thirty in the morning. He glanced at his watch then. He picked up the men with the rifle telescope. Two were asleep. He swung the sights up to her face and studied her again. A different look in the fading light around her. A face which in the darkness looked more like someone he knew. A sister. Someday a daughter. If he could have parted with it, the sapper would have left something there as his gesture. But he had his own faith after all.

I myself once experienced a display that in its resolute focus on an icon approximated the one just described. I had just checked into my hotel in Bombay, and it was apparent that something exceptional was astir. Asking what was afoot, I was told that at midnight the presiding deity of a temple four blocks away was to have her annual unveiling, which would last for the balance of that night. At the temple itself I learned the details. Though it would be a blessing to be anywhere in the temple that night, those who were in direct line of vision to the goddess when she was unveiled would receive her special *darshan*—the spiritual infusion that comes from being in the presence of the holy—and an especially auspicious year would be assured.

Happy in the good fortune of my timing, I returned to my hotel for a nap and late dinner, and at around ten o'clock left my room to head back to the temple. I told myself that I was doing this as a lay religious anthropologist, but I could also sense a touch of the polytheist stirring somewhere within me. I wanted my *darshan.*

I thought that two hours would suffice for me to find an appropriately aligned position in the temple, but I soon realized that my several years' absence had caused me to forget what India *is.* No sooner had I stepped from my hotel than I found myself in a river of humanity pouring toward the temple. Its density increased by the block, and by the time we were in sight of the temple I found myself wedged so tightly between people that I was actually lifted from the street. I thought of the press reports I had seen following major festivals in India—reports that carried statistics of the number of bodies that had been trampled to death. This was the one time in my life that I found myself fearing that my life might end in such an actuarial statistic.

My fear was unfounded. As so often happens in India, someone emerged out of nowhere and, seizing me by my hand, somehow managed to drag me out of the surging, hysterical mob. He led me down an alley for several blocks and then we doubled back toward the temple's farther side, where I found myself sitting on a rooftop with the veiled goddess in direct view. There we sat in silence (he spoke no English) awaiting the epiphany. On the stroke of twelve the veil was removed and—let's just leave it with this—I have no memory that the year that followed was *not* auspicious for me.

THE PRINCIPLE OF ONE-WAY MIRRORS

Halfway through my description of the four types seems like a good time to take a breath and dub in the principle of one-way mirrors, which I anticipated in the opening pages of this chapter. Polytheism leads into the subject particularly nicely because (as I have mentioned) the Monotheist's God often turns up in the background of the Polytheist's world.

The Polytheist accepts everything that the Atheist takes to be real and adds spirits to it. It is as if the Polytheist were to say to the Atheist, "I see everything you tell me you see. It is only what I see in addition and you do not see that divides us." To which the Atheist responds, "You mean what you *think* you see," for to the Atheist the Polytheist's additions are fictions. All of the disputes between the four types take this form, and I am proposing my one-way-mirror metaphor as a help in understanding the disputes between them. Looking upward, toward the heavens, the Atheist sees only mirror images of things in his own world. The spirits that the Polytheist situates on its far side, the Atheist dismisses as projections generated by unresolved conflicts in individuals and society—optical illusions, in other words. Meanwhile, when the Polytheist and Monotheist and Mystic look downward, they see not mirrors but plate glass. Or rather, they see only the things on the far side of the glass, for glass itself is invisible. To repeat: those who look downward see what is on both sides, whereas those who look upward see only what is below.

This difference pertains at every metaphysical frontier. Monotheists obviously do not discount the Polytheist's spirits; they merely baptize them, so to speak, converting them into angels and demons. Each successive echelon includes what is in the preceding ones and places it in a larger landscape. In the last resort, spiritual personality types are functions of how much each type perceives.

The Monotheist: There Is One God

Anthropologists often contrast Great Traditions with little traditions. Great Traditions are the historical, institutional religions that form the backbones of civilizations—Islam in the Middle East, Hinduism in India, etc.—and they are predominantly monotheistic. Invariably they are ringed with little traditions that are in many ways their opposites. Little traditions have no history. They have no buildings or institutionalized structure. Typically they center in a charismatic leader who appears to have remarkable powers, and they generally have difficulty outlasting her. (Women are conspicuous in little traditions, which is another way the latter differ from Great Traditions.) We are invited to think of oak trees surrounded by mushrooms at their base—and indeed, the ephemerality of little traditions has caused them to be nicknamed "mushroom sects."

Polytheism having already been covered, it is now monotheism's turn. The treatment can be brief because (sufficient for our purposes) the Monotheist's God—knowable and personal—was described in the preceding chapter. This frees us to attend here to the Monotheist's *relation* to that God. That it is an intimate, personal relationship goes without saying; Monotheists have no trouble at all in thinking of God as richly endowed with the finest qualities that human beings exemplify: wisdom, tenderness, mercy, compassion, creativity, love, and the like, which, elevated in degree, add up to glory. Love figures especially among these qualities. In the idiom of Hinduism's four *yogas*, *bhakti* and *karma yoga* are the natural routes to Ishwara or Bhagavan, two prominent names of the personal God in that tradition.

Krishna's love for the gopis (cow-herdesses), which was ardently returned, sets the tone in that tradition. In Vrindaban, Krishna's birthplace and the center of the Sri Caitanya Hari Krishna sect, I once heard a lecture that presented illicit love as the supreme model for our love for God. That startled me until the speaker's arguments set me straight. Moral categories do not apply in loving God, he cautioned first. He then explained that the aspect in illicit love that should be central in our love of God is its absolute, uncompromising character: illicit love—*love,* mind you, not sexual adventure—is uncomplicated and therefore wholehearted. This contrasts with married love, which always comes with obligations—to support a family, to remain faithful after time has tamped the flames of the novelty, and so on. Requited or not, illicit love—again, not crude illicit sex—is nothing if not romantic. We speak of being lovesick, of swooning with love. Our love of God should have that same passionate intensity that characterizes head-over-heels romance. One thinks of Dante and Beatrice, and Rumi and Shams of Tabriz.

Ethics enters as a corollary of passionate love when it is directed to God the creator, who "has the whole world in his hands." God loves the creatures she creates as if they were her children, so if we love God we will love them too. Ethics is absent from polytheism. It is inseparable from monotheism.

I have had occasion to remark that the idea of a personal God seems to give people more trouble now than it used to, so I will devote the remainder of this section to several glimpses of that concept at work.

- Novelist Anne Lamott says that her two favorite prayers are "Help me, help me, help me" and "Thank you, thank you, thank you."
- Malcolm Boyd, an Episcopal priest, wrote a book titled *Are You Running with Me, Jesus?* I have been surprised to find that supplication-in-question-form coming to mind when I find myself under stress.
- Several years ago when the American Academy of Religion was meeting in New Orleans, I took an evening off with a friend to visit Preservation Hall, a tourist attraction. It was the home of a Dixieland band that had formed around its pianist, affectionately known as Sweet Emma. Sweet Emma had suffered a stroke, but she was still at the piano, playing with her left hand while her right arm hung limply from her shoulder.

  There was standing room only, and we stood. The number that has stayed with me most clearly was this spiritual:

  O for a closer walk with Thee,
  Jesus, grant it if you please;
  Daily walking close to Thee:
  Let it be, Lord, let it be. . . .
  I am weak but Thou art strong,
  Jesus, keep me from all wrong;
  I'll be satisfied as long
  As I walk, Jesus, walk with Thee.

  You have to know that spiritual's melody to know what I am talking about here, and you have to hear the trombone laying down its melody with the suffering of three centuries of slavery behind the sound to understand the fervor in its supplication. When the last strains had died away and a thundering applause took over, my friend (whom I do not think of as particularly religious) turned to me and said, "They weren't *playing* that. They were *praying* it."

- The last instance that I shall mention also involves a personal anecdote, but one that runs deeper than the others that relate to me.

  I barely remember my maternal grandfather, who in the mid–nineteenth century went with his wife to China where they served as missionaries. He had retired and returned from China by the time I was born, but my mother shared stories of him. The most memorable account of him that she told me concerned a time when she was a child and bandit warlords were ravaging the countryside and pressing at the gates of the city they lived in. Word had arrived from the American Consulate in Shanghai that the Americans must evacuate the city immediately, but the escape routes were themselves dangerous. Just before they left their house (not knowing if they would ever return), she came upon her father in prayer. He was on his knees, his arms resting on the seat of the chair before him. His eyes were closed, and his face was turned upward. It was the expression it wore (she said) that she would carry with her to her grave. It was one of absolute trust.

The Mystic: There Is Only God

Value increases as we mount the four levels of existence. The atheist's world contains very little value, for value is an aspect of experience, and experience is in short supply in our fifteen-billion-light-years-across universe if organisms are its only seat. In sharp contrast, the polytheist's world teems with value. That increase is a mixed blessing, however, for it includes the value-opposites about evenly divided:

pain as well as pleasure, evil as well as good, and the host of other dualities. In the monotheist's world these dualities remain, but good has the upper hand. In the mystic's world evil drops from the picture and only good remains. There is only God.

This is a difficult notion to accept. Regularly it squares poorly with the morning news. (Once while Aldous Huxley was a visiting professor at MIT, I drove him to an evening lecture at Springfield College, which required an overnight stay before returning to Cambridge. Finding me reading the morning paper as I awaited his arrival for breakfast, he asked, "Has anything more than usually disastrous happened overnight?") There are, however, clues that help us understand it.

One helpful approach is to think of the value that can come to parts through their inclusion in larger wholes. Consider a magnificent painting. Block out all but a square inch of the canvas and the inch by itself does not amount to much. However, because the painting would not be what it is without it, the glory of the painting doubles back to dignify the inch in question.

It is the same with music. Taken in isolation, one tone is much like another. In the context of a great symphony, on the other hand, a given note acquires grandeur for being exactly the right note in the right place. The symphony's perfection depends on it, and because of that fact, the symphony in effect beatifies the note (if this is not speaking too fancifully). Thank you very much, B-flat from the oboe. You were needed. In "The Red Wheelbarrow," William Carlos Williams transfers this point to the visual field.

So much depends on
A red wheelbarrow
Glazed in rain water
Beside the white chickens . . .

There is a second step in this line of thinking—a step that Plato develops in his *Phaedrus,* where he describes the ascent from loving beautiful bodies, through loving beautiful souls, to loving beauty itself. To continue with music, think of the most memorable concert you ever attended, so flawless that it climaxed in what critics refer to as "aesthetic seizure." It took time for that state to arrive, but on the evening in question the parts that went into its making—instruments, players, the featured number on the program, the director who brought the whole evening together—worked so perfectly "in concert," as we say, that at some point multiplicity fell away entirely. Before that point, you had followed the symphony through its familiar themes and transitions, but when the magic kicked in, you, the listener, lost track of the fact that it was *you* who were hearing this glory. Wave after wave of the music was all there was. For purposes at hand, there was only God.

Bring, now, the God in that aesthetic seizure into direct line of vision. Sufis are famous for speaking of God-intoxication. In extreme instances of that state, the *dervish* can lose awareness of himself to the point of becoming dissociated in the psychological sense of that word—in other words, he no longer knows who or where he is or what he is doing.

I witnessed this once during a Sufi gathering in Tehran. The hour had grown late. For the climax of the ritual the few candles that had dimly lit the hall were extinguished, and a hypnotic, pounding chanting took over. Gradually I became aware of a man who was sitting opposite me in the swaying circle, his form dimly silhouetted against the faint light from a transom behind him. His movements grew erratic and then agitated as they broke with the swaying rhythm of the circle; then, after a minute or two, convulsions set in, punctuated with loud outbursts of "Allah, Allah, Allah." Quickly two "bouncers" (as I found myself thinking of them later, because of their hefty size) appeared behind him, smothered him in their embrace, and held him pinioned until his seizures subsided. In ecstatic moments such as that, it could be literally the case that there was only God in what that *dervish* experienced. Allah could very well have filled his entire mental horizon.

Sufis respect their ecstatics, referring to them affectionately as spiritual drunkards who hang out in God's tavern; but they hold in higher regard those who can see God everywhere while they are sober—which is to say, see God everywhere in daily life. This requires considerable reflective talent, though we must never forget that in matters spiritual, thinking comes closer to seeing than to reasoning. Reasoning brings indirect knowledge (knowledge *about*), whereas intuition brings direct knowledge (knowledge *of*). The latter causes thoughts to circle their objects, spiraling around them conically until in a flash of insight they penetrate their object like a drill.

Here the circlings take the form of the reflections that I presented while discussing the Godhead in the preceding chapter. To be infinite, God must include all possibilities. Finitude is possible—here we are as witnesses—so finitude must be included in God, together with all its gradations. That sounds like a syllogism, but if it remains at the level of logic only it will speak to no one. Only if the point is grasped *intuitively* will it become religiously effective. Then every moment is recognized as being God in this particular mode of his/her/its veiling. Secularists see only the veil; those with religious sensibilities glimpse God through the veil; mystics see only God, because they realize that the veil is necessary to God's being God and therefore is a part of God. This does not cause mystics to disregard the veil. Indeed, at times they experience it as so thick that it causes them to cry out, "My God, my God, why hast thou forsaken me?" But in their heart of hearts they understand that God is fully present everywhere and in everything and that his seeming absence is required if he is to share his infinity while remaining in himself the absolute perfection that he is. That perfection prevails. God is all in all.

Ram Dass tells of walking with his guru, Neem Haroli Baba, in Bangladesh. The suffering around them was so heart-wrenching that he could hardly stand it. His guru kept saying, "Can you see how perfect it is?"

As was just suggested, the great problem for mystics is the corollary that attends God's being all in all, which is that there is no evil. Theodicy, which wrestles with the problem of evil, is the Gibraltar on which every rationalistic system eventually founders, so I must deal with it here, though briefly. I shall limit myself to two short suggestions.

1. If a two-year-old drops her ice-cream cone, that tragedy is the end of the world for her. Her mother knows that this is not the case. Can there be an understanding of life so staggering in its immensity that, in comparison to it, even gulags and the Holocaust seem like dropped ice-cream cones?

2. The only professional athlete I have known is a retired football player who played linebacker for several professional football teams, including the Los Angeles Dons, during the 1940s. One Sunday—he had already been retired for many years at this point—he took me for brunch to the Los Angeles Athletic Club (where he had been athletic director and was still senior vice president). I found myself inquiring about the injuries and operations he had suffered in the course of his athletic career. His catalogue was long, and the items on it had left him with pains that would remain with him for the rest of his life. But when I asked him if it had been worth it, he seemed surprised by the question. "Of course," he said. "I was so fortunate. So much that I have learned and enjoyed about life came out of my athletic career."

CHAPTER 16

SPIRIT

At one point in Barbara Walters's two-hour-long interview with Monica Lewinsky, Walters quoted President Clinton as confessing that he had sinned in his relationship with Lewinsky, and she asked if Lewinsky thought she too had sinned. Lewinsky appeared taken aback. She hesitated, shifted in her chair, and then answered, "I'm not very religious. I'm more spiritual."

That answer points to a problem in our collective thinking about religion: a cloud has descended over the very word itself. (The case parallels the previously discussed case of the word *hierarchy*.) Uncontaminated, *religion* is a noble word; deriving as it does from the Latin *religio,* to rebind, the word targets what religion is essentially about. But because it challenges the prevailing worldview, it has lost some of its respectability. Mention the word in public and its sins are what jump first to mind. Still, it is difficult to argue that religion has nothing to be said for it, which leaves us with Tonto's remark when, on entering a barn with the Lone Ranger, he took several good sniffs and pronounced, "There's got to be a horse in here somewhere." Enter the word *spirituality* to name (without specification) what is good about religion. Being no more than a human attribute, spirituality is not institutionalized, and this exempts it from the problems that inevitably attend institutions—

notably (in religious institutions) the in-group/out-group tensions they tend to breed.

This point was touched on in the chapter on higher education, so I need add only a single point to what was said there. It is a bad sign when *spiritual,* an adjective, gets turned into a noun, *spirituality,* for this has a dog chasing its own tail. Grammatically, *spirit* is the noun in question, and *spiritual* its adjective. *Spirituality* is a neologism that has come into existence because *spirit* has no referent in science's world, and without grounding there, we are left unsure as to what the word denotes.

This chapter seeks to resolve that uncertainty, and I will phase into the task by way of the *New Yorker* magazine. One of the pleasing features of that magazine over the years has been its running lampoon of religiosity through cartoons and squibs that puncture piosity. The item I have in mind reported a sermon title on a church bulletin board (with town and church identified) that read, "Live Offensively." The *New Yorker*'s comment was, "No day is too short to be just a little repulsive." That quip comes to mind here because in this final chapter—it is my last chance—I intend to seize the offensive and take my chances on how that comes across.

The Self/World Divide

As was noted in Chapter 11, ever since Descartes split the world into mind and matter (or subject and object), philosophers and scientists have tried to bridge the chasm without success. I will confine myself to a single example, the psychology of perception.

If we try to connect an animal in the wild to its environment via what textbooks say about the physiology of perception—breaking the act down into neural components that must then be hooked together—we encounter so many inexplicable gaps that rationally we would have to conclude that the animal does not perceive its world at all. Yet all the while it *behaves* as if it perceives the world and proceeds toward food and shelter almost unerringly. With J. J. Gibson's *The Ecological Approach to Visual Perception* pointing the way, animal psychologists are coming to see that they have lost sight of this incontestable fact. Trying to account for knowledge as inference from noetic bits does not work. We must begin the other way around, with the recognition that there is a world out there and that animals are oriented to it.

With that running start toward Spirit, I advance from animals to human beings.

Tacit Knowing

I begin with a human ability that baffles epistemologists completely, one that differs markedly from reason. Reason performs logical operations on information that is in full view and can be described and defined. Again and again, though, we find that our understanding floats on operations that are mysterious because all that we seem able to know about them is that we have no idea how they work. We have hunches that pay off. Or we find that we know what to do in complicated situations without being able to explain exactly how we know. The knowledge in question is unconscious, yet it enables us to perform enormously complicated tasks, from reading and writing to farming and composing music. Expertise is coming to be recognized as more intuitive than cognitive psychologists had suspected. These students of learning are finding that when faced with exceptionally subtle tasks, people who "feel" their way through them are more creative than those who consciously try to think their way through.

This explains why computer programmers no less than psychologists have had trouble getting the experts in their field to articulate the rules they follow. The experts do not follow rules. This bears crucially on artificial intelligence, whose theorists are reluctantly coming to see that machines can never replicate human intelligence because we are not ourselves thinking machines. Each of us has, and uses in every moment of the day, a power of intuitive intelligence that enables us to understand, to speak, and to cope skillfully with our everyday environment. Somehow that intuition summarizes everything we have ever experienced and done, and enables that summary to shape our present decisions.

That states the matter, but abstractly, so we need an example to give it force. Japanese chicken-sexers are able to decide with 99 percent accuracy the sex of a newborn chick, even though the genitalia are not visually distinguishable. No analytic approach to learning the art could ever approach such accuracy. Aspiring chicken-sexers learn only by looking over the shoulders of experienced workers, who cannot explain how they themselves do it. Exposed to the art, the novices "get the hang of it," as we say.

Talking parrots provide an even more startling instance of the obscure talent I am tracking. What goes on when a parrot imitates the voice of its owner, or the bark of a dog, or human laughter? Presumably, the parrot has some sort of conscious life. It hears the voice, it hears the bark or the laughter, and presumably it wishes (in a way that rudimentarily corresponds to our desire to do something) to imitate the sound.

But then what happens? When you think of it, it is one of the most extraordinary things you can imagine. Something incomparably more intelligent than the parrot itself sets to work and proceeds to activate a series of sound organs that are totally different from those of human beings. People have teeth, a soft palate, and a flat tongue, and the parrot has no teeth, a rough tongue, and a beak. From these, however, it proceeds to organize its absolutely different apparatus to reproduce words and laughter—so exactly that we are very often deceived by it into thinking that what is in fact the parrot talking is the person herself making the utterance. The more we reflect on this, the stranger it becomes, for in the course of evolutionary history parrots have not been imitating human beings from time immemorial; people arrived after the parrots' adaptive mechanisms were in place. We have here an *ad hoc* piece of intelligent action, carried by some form of intelligence within the parrot, that cannot be explained by evolutionary conditioning.

We find these examples astonishing, but the talent it jolts us into noticing is *in kind* one that directs every step of our lives. What we call *prudence* provides an everyday example. Functioning in something of the manner of a hidden gyroscope, it monitors our inclinations and comes up with a yes or a no, the two magical words of the will. To do this, it spins no theories. Instead, it synthesizes all we have learned and brings this synthesis to every decision we make. In doing so it provides dozens of answers to dozens of questions, and—because it gives no evidence of caring about their mutual agreement—conveys the impression that each particular answer is absolutely *ad hoc*. This gives the talent the air of practical poetry, for each particular answer arises spontaneously while being for the most part appropriate and (for the moment in question) conclusive. The spontaneity of prudence is deceptive, however, for if we reflect on the matter we find that all its *ad hoc* answers arise from a whole that directs them and makes them appropriate; its activities are prodigiously married. The integral truth of our being, from which it springs, envelops and inspires everything we consciously and unconsciously do, giving our lives their form and style, and seeing to it that each action and decision reflects that style.

Nothing I have thus far said is new. For some time chemist-turned-philosopher-Michael Polanyi, evolutionary biologists, and developmental psychologists have been talking about tacit knowledge—cognitive underpinnings that are indispensable to our knowing but that operate unconsciously. All of these investigators, however, assume that mental operations that we cannot explain ride the waves of simpler operations that are rationally intelligible. In short, they assume that the *more* derives from the *less*. Traditionalists assume the opposite, and from this single difference the two worldviews that this book has been juggling separate as day from night.

Defiantly I have taken my stand with the traditionalists, and in this closing chapter, as I said, I am taking the initiative. I want to try to drag modern investigators kicking and screaming toward the possibility that, given the way they have set things up, they cannot get to where they want to go—as computer programmers say, garbage in, garbage out. Smashed to smithereens, Humpty Dumpty cannot be put together again—attempts to do so are Kafka's cage searching for a bird. This suggests that it might be sensible to go back to where Humpty Dumpty was sitting, happy and whole on the wall. Wholeness comes first, multiplicity later; the many derive from the one.

Approaching the matter from this direction will not lessen the mystery of the progressions involved and may not hold many suggestions for scientific research, though maverick biologist Rupert Sheldrake is toying with some possibilities here. It might, however, carry suggestions for how we should live. That would be no small benefit, given Richard Rorty's observation that the legacy of Descartes's dualism has been to cause philosophers to replace the search for wisdom with the search for certainty, and to turn toward science rather than toward helping people attain peace of mind.

Spirit and Its Outworkings

The wholeness with which traditionalists begin is God: "Hear O Israel, the Lord our God, the Lord is one." When capitalized, Spirit is a synonym for God, and I am using it as such here with emphasis directed toward God's presence and agency in the world, as when in Genesis "the Spirit of God moved on the face of the waters" and in human beings the Holy Spirit is God's workings within them. Spirit in this chapter is the *imago dei* of Jews and Christians, the *Atman* of the Hindus, the Buddha-nature of the Buddhists, the Uncarved Block of East Asia, and the "best stature" in which the Koran tells us human beings have been made—Figure 2 depicts this diagramatically. Whether Spirit, thus conceived, is identical with God or God's mirror image is negotiable. Mystics champion identity; monotheists insist that a distinction remains.

Having in this chapter approached Spirit through the back door—by way of the difficulties that we face in trying to understand in its absence how human knowing works—I turn now to indicating how things might look when Spirit is taken to be fundamental to the world. (Is there any *reason* for thinking that consciousness, or sentience, or awareness—all of these being names for the point where Spirit first comes to human attention—is less fundamental than matter? That we can lay our hands on matter but not consciousness is not a reason.)

I begin with what Plato would have called a likely tale. What if, in the Big Bang, it was Infinite Omniscience that exploded? According to the traditional law of inversion, what is logically prior arrives last in point of time. Here that translates into God's being both the causal beginning of things and their temporal end. From God we derive, and to God we eventually return.

Chronologically, the sequence begins with the meagerest possible existences that become increasingly complex as time proceeds. But note that in this scenario intelligence is present in those microscopic entities at the very start—there is a Buddha in every grain of sand. In the early scientific view atoms were governed by laws they had no part in devising, but with indeterminacy's entrance into the picture—Heisenberg's uncertainty principle—laws are now statistical averages of the way atoms decide to behave. Scientists read "decide" here metaphorically; though not all scientists, for Freeman Dyson writes that "it appears that mind, as manifest by the capacity to make choices, is to some extent inherent in every atom." His opinion hasn't entered the textbooks, but it agrees with tradition where sentience pervades.

And though in the smallest things God's omnipresent omniscience is veiled under the thickest conceivable veil, the tiniest bit of sentience that surfaces in those things is *of a kind* with omniscience and is backed by it. Why do not particles content themselves with being just what they are—particles? Whence comes this drive toward complexity that leads (on the planet we know firsthand) to plants, animals, and rationality? Because intelligence is actively working to free itself from its stifling veils and give itself more elbow room for movement in the finite world. *That* is why tacit knowledge comes together and serves us so well. Its components (under the final direction of the omniscience that orchestrates everything) are up to something, that "something" being their working for the greater largess just mentioned. The same with the biological bits that enable parrots to talk, and right on up to ourselves, we human beings. Reflecting on the fact that under hypnosis our body can rid itself of warts, Lewis Thomas writes: "There almost has to be a Person in charge, running matters of meticulous detail beyond anyone's comprehension, a skilled engineer and manager, a chief executive officer, the head of the whole place. I never thought before that I possessed such a tenant. Or perhaps more accurately, such a landlord, since I would be, if this is in fact the situation, nothing more than a lodger." Lewis's CEO is my Spirit, causing ingredient elements everywhere to reenforce one another in ways that "make sense" in the literal meaning of manufacturing sense where previously there was none.

As for the self/world divide, it is a Cartesian conceit. Self-contained atoms in the void that must collide to connect went out with Greek proto-science long ago; field theory now reigns. *Pratitya-samutpada* (interdependent arising). Indra's net in which every jewel in the net reflects every other jewel and the reflections *in* every other jewel. David Bohm's implicate order. This way of viewing things tells us nothing more about the details of the process at work than the analytic approach does, but it offers a *kind* of explanation that is more intelligible than its alternative. When an adult solves a riddle or laughs at a joke, it is no surprise, for the *capacity* to do so was in place. For a child who has not yet acquired the requisite capacity, no amount of explaining—piecing together of linguistic bits—does the trick.

The issue in question—more from less or less from more—hits us in the face when we consider how we got here. Darwinists consider it a proven fact that novel qualities—life, sentience, and self-consciousness—can derive from the rearrangement of elements that themselves lack those qualities. The explanation that is offered for how these rabbits appear out of hats is to say that they emerge. What that explanation overlooks is that "emergence" is a descriptive, not an explanatory, concept. It explains nothing.

Consciousness and Light

From Shankara, Ramanuja, the Abhidharma, and the Madhyamika in Asia, to the magisterial writings of Augustine, Plotinus, Thomas

Aquinas, Avicenna, Averroes, and Ibn 'Arabi in the West, the less-from-more worldview has been worked out with a precision and detail that rivals its scientific opposite, but of course that is not for here. Here, before bringing this book to a close, I want only to consider for several paragraphs a single step in the sequence from Spirit to matter in order to suggest how it escapes the problems that attend Descartes's self-world divide by positing a single source for them both. (Descartes himself was traditional enough to posit God as the source of *res extensa* and *res cogitans,* but as has been said, philosophers have ditched that source.) I have not bothered the reader with proof-texting my assertion that traditional philosophers did not work from the premise of a subject-object split, but given the importance of the point it might be well to provide at least one example. Hilary Armstrong tells us that for Plotinus, the Intellect (a technical term) "is the level of intuitive thought that is identical with its object and does not see it as in some sense external."

We should not conclude from the identity they worked from that traditional philosophers were blind to distinctions. Obviously, our inner lives and the world in which they are set are different in certain ways, but they derive from a common source. Think of an inverted V. Its apex is Spirit, and the two arms that reach down from it are consciousness (or more inclusively, sentience) and matter. This section tracks their relationship.

If consciousness is not simply an emergent property of life, as science assumes, but is instead the initial glimpse we have of Spirit, we ought to stop wasting our time trying to explain how it derives from matter and turn our attention to consciousness itself. The image on a television screen provides an analogy for what we then find. The television lights up its screen, and the film in the video we are watching modifies that light so as to produce any one of an infinite number of images. These images are like the perceptions, sensations, dreams, memories, thoughts, and feelings that we consciously experience—we might think of them as the *contents* of consciousness. The light itself, without which no images would be possible, corresponds to pure consciousness. We know that the images on the screen are composed of this light, but we are not usually aware of the light itself. Our attention is caught up in the images that appear and the stories they tell. In much the same way, we know we are conscious, but normally we are aware only of the many different experiences, thoughts, and feelings that consciousness presents us with. Consciousness proper—pure consciousness, consciousness with no images imposed upon it—is the common property of us all. When (in introspection or meditation) we detect *pure* consciousness, we have every reason to think that what I experience is identical with what you experience in that state. And identical with what God too experiences, not in degree but in kind. For at that level, we are down to what consciousness *is,* namely infinite potential—receptive to any content that might be imposed on it. The infinitude of *our* consciousness is only potential whereas God's consciousness is actual—God experiences every possibility timelessly—but the point here is that our consciousnesses themselves are in fact identical.

That is the left, subjective, arm of the inverted V. The right descending arm represents Spirit branching out to create the physical universe. Its instrumentality for doing this is light, or as scientists say, photons. (If I try to move to what might be beyond or behind or beneath photons—a strict impossibility in my case—a no-man's-land opens up where nobody really knows what goes on.) Photons are transitional from Spirit to matter, because (as we saw in the chapter on "Light") they are only quasi-material while producing things that are fully material. Scientists would give their eye-teeth to know what the non-material component of photons is. For religionists, it is Spirit.

Notice the parallel with consciousness here. All that we typically see, optically, is light that is overlaid with images of one sort or another. The photons that strike the optic nerve of the eye are known only through the energy they release, which energy produces in us the sensation of light. *That* light, though, is a quality of mind, for to repeat, we never see photons, which is to say light in the form in which it pervades the objective world. But the light that we see and the photons in the objective world derive from the same source and carry that the trace of that source—Spirit—within them.

In some such way as this, traditionalists see physics affirming with *Genesis* that in the beginning there was light. And (as again we saw in the chapter on Light) there continues to be light, for light underlies every process of nature, wherever and whenever. Every exchange of energy between atoms involves the exchange of photons. Every interaction in the material world is mediated by light; light penetrates and interconnects the entire cosmos. "An oft-quoted phrase comes to mind," physicist-turned-metaphysician Peter Russell remarks: "God is Light. God is said to be absolute—and in physics, so is light. God lies beyond the manifest world of matter, shape, and form, beyond both space and time—so does light. God cannot be known directly—nor [as photons] can light." When on the religious side we think of St. John's reference to "the true light, which lighteth every man that cometh into the world," and the *Tibetan Book of the Great Liberation*'s reference to "the self-originated Clear Light of the Void, eternally unborn, shining forth within one's own mind," the correlation is remarkable. Reinforce it with this word from the Islamic tradition. Abu'l-Hosain al-Nuri experienced light "gleaming in the Unseen. I gazed at it continually, until the time came when I had wholly become that light."

Happy Ending

In contrasting the great outdoors with the tunnel in the first half of this book, I noted that the religious worldview conforms to the most successful plot device ever conceived—namely, a happy ending that blossoms from difficulties necessarily confronted and overcome. Thus far I have not given content to that ending, but the time has come to do so.

In the scientific worldview, matter—its foundation—cannot be destroyed; it changes its form but never disappears. The same holds for consciousness when it replaces matter as foundational. How consciousness changes when it "drops the body," as Indians say, is the great unknown, but Ruth Ann in Barbara Kingsolver's *The Poisonwood Bible* points toward the religious answer. Having died as a child in the Congo she has assumed the form of a serpent in keeping with Congolese beliefs, and as a green snake lying on a tree limb she is watching her mother and sisters who after many years have returned to Africa to search for her grave. What she wishes she could tell them is, "Listen: Being dead is not worse that being alive. It is different, though. You could say the view is larger."

Charles Tart, professor of psychology at the University of California, Davis, agrees with Ruth Ann. Tart is one of the two academics I know who has devoted his career to studying paranormal manifestations of consciousness—near-death experiences, telepathy, clairvoyance, precognition, psychokenesis, seances, and the like—and I heard someone ask him if he thought his consciousness would survive his bodily death. He said he was certain that it would, but added that he hadn't a clue as to whether he would then recognize it as *his* consciousness.

Religions teach that after death one *is* aware of who one has been and is, and add that one's work is at that point not completed.

Those who teach reincarnation hold that the soul returns to earth to take up its unfinished business here. (It makes an exception for *jivamuktas*—souls that have achieved liberation while still embodied—but those are extremely rare.) How many rounds are required to complete life's agenda depends on how adept the soul is at learning life's lessons.

The alternative to reincarnation locates what remains to be done on a different plane of existence. The Abrahamic religions stand together in doing this, though Judaism, Christianity, and Islam contain minorities that subscribe to reincarnation. (As one example, we find Rumi saying, "I died as mineral and became a vegetable. I died a vegetable and became an animal. I died an animal and became human. When was I less by dying?") Tibetans would call the place where (in the official Western version) the remaining business is transacted a *bardo. Purgatory* is one name for it among Westerners, as is *hell* (a notion to which I shall return).

As for what the remaining business *is,* it is the cleansing that must be accomplished before the soul can enter an abode of total purity, variously designated as the Pure Land, Happy Hunting Grounds, Heaven, the Western Paradise, and others. Fire is typically cited as the cleansing agent. Some accounts take the word literally while others employ it metaphorically. The Koran includes both readings. Literalism predominates; but Sufis take refuge in the verse that reads, "We have hung every man's actions around his neck, and on the Day of Judgment a wide-open book will be laid before him." What death burns away (these Sufis say) is self-serving rationalizations and defenses. When they are gone, the soul will be forced to see with total objectivity how it has lived its life. In the uncompromising light of that vision, where no dark and hidden corners are allowed, it is one's own actions that rise up to accuse or confirm. Once the self is extracted from the realm of lies, the falsities with which it armored itself become like flames, and the life it there led like a shirt of Nessus.

When hell is understood as a cleansing station that one exits when the cleansing is complete, it fits with the above-mentioned notions of *bardo,* purgatory, and (I now add) hell as a temporary abode. Eternal damnation is another matter, and I will approach it by way of an anecdote.

It was 1964 and I was using a semester's leave to continue my researches in India. At the moment to be described, I was conversing with one of a number of gurus whose reputations had taken me to the foothills of the Himalayas, when suddenly there appeared in the doorway of the bungalow I was in a figure so striking that for a moment I thought I might be seeing an apparition. Tall, dressed in a white gown, and with a full beard, it was a man I came to know as Father Lazarus, a missionary of the Eastern Orthodox Church who had spent the last twenty years in India. Ten minutes after I was introduced to him I had forgotten my gurus completely—he was much more interesting than they were—and for a solid week we tramped the Himalayan foothills talking nonstop.

The reason I am relating this is for one particular exchange in the week's conversation. I had told him that I found myself strongly attracted to Hinduism because of its doctrine of universal salvation. Everyone makes it in the end. Its alternative, eternal damnation, struck me as monstrous doctrine that I could not accept.

Brother Lazarus responded by telling me his views on that matter. They took off from the passage in Second Corinthians where Saint Paul tells of knowing someone who twelve years earlier had been caught up into the third heaven, whether in the body or out of the body he did not know. For emphasis, Paul repeats that last point—"whether in the body or out of it I do not know"—before he goes on to say that in that heaven the man "heard things that were not to be told, that no mortal is permitted to repeat." Father Lazarus quoted the passages verbatim. Paul was speaking of himself, Father Lazarus was convinced, and the secret he was told in the third heaven was that ultimately everyone is saved. That is the fact of the matter, Father Lazarus believed, but it must not be told because the uncomprehending would take it as a license for irresponsibility. If they are going to be saved eventually, why bother?

That exegesis solved my problem and has stayed in place ever since. A number of years later I was pleased to find it confirmed by Sufis who (likewise quietly) accept at face value the verse in the Koran that reads, "Unto Him all things return."

On the heels of this esoteric/exoteric dispute over whether we are all saved, there is another. At the end of our journey do we merge with the Godhead or enjoy the beatific vision of God forever? Monotheists champion the latter, mystics the former. Ramakrishna, who had a genius for embracing both horns of a dilemma, identifying with both sides, exclaimed in one of his monotheistic moods, "I want to taste sugar; I don't want to be sugar." The standard metaphor for the mystics' alternative is: the dewdrop slips into the shining sea.

Having allowed himself the right to his own opinion on one important theological doctrine, Father Lazarus would not likely deny me my right to exercise *my* private conscience on another. It is this. We are, I believe, allowed to choose between the alternatives just presented. Resorting again to a likely tale, this is the way I envision matters. Stated in the first-person singular, my extravagantly explicit scenario is this:

After I shed my body, I will continue to be conscious of the life I have lived and the people who remain on earth. Sooner or later, however, there will come a time when no one alive will have heard of Huston Smith, let alone have known him, whereupon there ceases to be any point in my hanging around. Echoing John Chrysostom's reported farewell, "Thanks, thanks, for everything; praise, praise, for it all," I will then turn my back on planet earth and attend to what is more interesting, the beatific vision. As long as I continue to be involved with my individuality, I will retain the awareness that it is I, Huston Smith, who is enjoying that vision; and as long as I want to continue in that awareness I will be able to do so. For me, though—mystic that I am by temperament, though here below not by attainment—after oscillating back and forth between enjoying the sunset and enjoying Huston-Smith-enjoying-the-sunset, I expect to find the uncompromised sunset more absorbing. The string will have been cut. The bird will be free.

EPILOGUE

We Could be Siblings Yet

Midway through his book *Gandhi's Truth,* Erik Erikson interrupts his psychoanalytic study of Mahatma Gandhi with a very different kind of chapter—a short chapter that he titles "A Personal Word." It takes the form of a letter addressed to Gandhi as if he were still alive. Its salutation, Mahatmaji (Revered Great Soul), makes clear Erikson's deep regard for Gandhi, but Erickson then explains that he wants to tell him in person, as it were, that the second half of his book will call attention to some flaws in Gandhi's character that Erikson thinks his psychoanalytic training has enabled him to detect. He feels confident, he says, that Gandhi would want him to report what he has found, for no one dedicated his life more unswervingly to truth than did Gandhi.

The sensitivity with which Erikson leads into his approaching criticisms makes his letter a model for this Epilogue, whose title readers will recognize as paraphrased from Chief Seattle's most famous oration. Paraphrasing Erikson, I proceed as follows:

Respected scientists, and by extension keepers of the high culture—but let me stop right there, for that salutation is poorly phrased. It should read, "Respected keepers of our high culture with a handful of polemical scientific materialists thrown in," for the majority of scientists are sensitive and tolerant citizens who treat religion with respect, just as the great majority of religious people belong to moderate denominations that treat science with respect. Dogmatic scientific materialists are as exceptional as dogmatic religious fanatics, but because they stir things up and the media loves a good fight, their number and importance gets exaggerated. So I shall rephrase my salutation and address my "letter" generally to the officiants of our high culture while aiming it pointedly at the sprinkling of militant scientists who make up in polemical zeal what they lack in numbers. With that correction I shall start over and try to say it right this time.

Respected adversaries, I should like to suggest what is required of militant scientists if the two great shaping forces of history are to join hands in the coming century. Like a dysfunctional family trying to become functional, the change will take time, but as a step toward it, I suggest that you try to understand where we believers are coming from.

The polemical among you are not good at doing that. My shelf of books on science-for-the-laity is as long as my shelves on each of the major world religions, but I will be very much surprised if you can say as much from your side. Your standard criticisms of religion sound so much like satires of third-grade Sunday school teachings that they make me want to ask when you last read a theological treatise and what its title was.

The greats in your number did better than you in this regard. Schrödinger's trilogy of small books for the general public ends with the Upanishads' resounding refrain, "Atman is Brahman." Niels Bohr credited Søren Kierkegaard's writings with sparking his doctrine of complementarity. Robert Oppenheimer read the Bhagavad Gita in Sanskrit (which I cannot) and quoted from it when the first atomic bomb exploded in New Mexico. Werner

Heisenberg and Arthur Compton were the leading lights in the conference on Science and Human Responsibility that Washington University mounted in the 1950s, and its program reserved Sunday morning for divine worship.

What I feel that you do not understand is why we, your potential partners, are so persistent in pressing our case. You know by heart the pathological reasons for our doing so, but the first requirement in conflict resolution is to try to understand the person on the opposite side of the negotiating table at his or her best. At our best, we seem to possess a sensibility that you lack, and I shall try to describe it.

Most simply stated, to be religiously "musical" (as Max Weber confessed he was not) is to possess a distinctive sensibility that I shall call the "religious sense." It has four parts that lock together into a single whole.

1. The religious sense recognizes instinctively that the ultimate questions human beings ask—What is the meaning of existence? Why are there pain and death? Why, in the end, is life worth living? What does reality consist of and what is its object?—are the defining essence of our humanity. They are not just speculative imponderables that certain people of inquisitive bent get around to asking after they have attended to the serious business of working out strategies for survival. They are the determining substance of what makes human beings human. This religious definition of human beings delves deeper than Aristotle's definition of man as a rational animal. In the religious definition, man is the animal whose rationality leads him to ask ultimate questions of the sort just mentioned. It is the intrusion of these questions into our consciousness that tells us most precisely and definitively the kind of creature we are. Our humanness flourishes to the extent that we steep ourselves in these

questions—ponder them, circle them, obsess over them, and in the end allow the obsession to consume us.

2. Following on the heels of the above, the religious sense is visited by a desperate, at times frightening, realization of the distance between these questions and their answers. As the urgency of the questions increases, we see with alarming finality that our finitude precludes all possibility of our answering them.

3. The conviction that the questions *have* answers never wavers, however, and this keeps us from giving up on them. Though final answers are unattainable, we can advance toward them as we advance toward horizons that recede with our every step. In our faltering steps toward the horizon we need all the help we can get, so we school ourselves to the myriad of seekers who have pondered the ultimate questions before us. You scientists learn from your precursors too; Isaac Newton expressed this nobly when he said that the reason he saw further than his predecessors was because he stood on their shoulders. But it is easier in science to see what should be retained and what retired, for scientific truths are cumulative whereas religious truth is not. This requires that we keep dialoguing with our past as seriously as this book has tried to do, while also dialoguing expectantly with our present (as this book has also tried to do.) Karl Barth used to say that he faced each day with the Bible in one hand and the morning newspaper in the other.

4. Finally, we conduct our search together—collectively, in congregations as you do in your laboratories and professional societies. Emile Durckheim, the nineteenth century sociologist, thought religion was entirely a social affair, a reification of the shared values of the tribe. Today our individualistic society comes close to assuming the opposite, that religion is altogether an individual affair. (Charles

Taylor criticizes William James's *Varieties of Religious Experience* on this count.) As usual the Buddha walked the middle path. "Be ye lamps unto yourselves," for sure; but do not forget that the *sangha* (the monastic community, and by extension the company of the holy) is one of the Three Jewels of Buddhism.

As I try to describe the religious sense, my mind goes back to a night when I felt it working in me with exceptional force. My wife and I were spending a week in the dead of winter in Death Valley, California, and on the full-moon night that we were there I awoke around two A.M. to a call that seemed to come from the night itself, a call so compelling that it was almost audible. Hurrying into some clothes, I answered it. Stepping out of doors, I found that not a breath of air was stirring. The sky held no clouds to conceal the panoply of stars ascending from the circling horizon. It was one of those totally magical nights and moments.

For half an hour or so I walked the road, without (as I remember the epiphany) a thought in my head. It may have been as close as I have ever come to the empty mind that Buddhists work toward for years.

There my powers of description shut down, so I was happy a year or two later to come upon this poem by Giacomo Leopardi which, on reading, I recognized as giving words to the night in question. In that poem, a nomadic shepherd in Asia is posing questions to a moon that seems to dominate the infinity of earth and heaven—questions whose horizons are themselves infinite:

And when I gaze upon you,
Who mutely stand above the desert plains
Which heaven with its far circle but confines,
Or often, when I see you
Following step by step my flock and me,

Or watch the stars that shine there in the sky,
Musing, I say within me:
"Wherefore those many lights,
That boundless atmosphere,
And infinite calm sky? And what the meaning
Of this vast solitude? And what am I?

INDEX